Algebra & Trigonometry

Electrical and Electronic Engineering Design Series

Electric Circuits Analysis and Design

Electronic Circuit Design with Bipolar and MOS Transistors

CMOS Circuit Design Analog, Digital, IC Layout

Digital Design Logic, Memory, Computers

Analog Filter Design

Error Correction Code Design

Electronic Circuits – Practical Learning

Computer Science Design Series

Programming with MFC & Visual STUDIO

Learn to Program from Scratch with MFC and the C languages

Mathematics

Arithmetic – Integers, Fractions, Decimals

Algebra & Trigonometry

Mathematics Beyond the Calculus

Algebra and Trigonometry

Nicholas L. Pappas. Ph.D.

ISBN 9781713290957

A Message about this Text: The subject is essentially endless. The purpose here is to say enough about the subject so that you, the reader, have a running start when you apply this knowledge to your work.

Prerequisites: Competence in Arithmetic

We believe important benefits accrue by doing the problems carefully, and by formulating the equations if only to copy them. These efforts in effect provide "startup" experience.

Once you have some experience we are confident that you will be able to expand your know how with reasonable effort.

A Message from the Author: I have worked continuously in the electronics industry since 1950 except for 11 semesters teaching at San Jose State University (Professor and Chair Computer Engineering 1988-1993). There I discovered my talent for teaching such as it may be. After War2 I attended Lehigh University, and then transferred to Stanford where I earned the MS degree and, while working at HP in the early 1950's, the Ph.D. EE degree. (Somehow I did not get the word and formally apply for the BS degree.) Hardware design has been my principal activity. I have learned enough about assembly language, Forth, C, and C++ to design the software I need for my projects.

Preface

Algebra
This is about the fundamental ideas of Algebra, understanding why and how Algebra works.

A focus on general methods for solving algebraic equations allows one to know how to solve any problem. The numerous special methods have limited value.

Sometimes an equation is not in the desired form. Algebraic operations are used modify the form of the equation by making the *same* changes to both sides of =. The equality is not upset if the *same* changes are made to both sides of =.

A polynomial in one variable x is defined and its essential properties are discussed. The + × – ÷ operations on polynomials are explained.

The Remainder Theorem is explained.

Methods finding factors of polynomials and Newton's method for finding zeros of polynomial (factors) are presented.

To solve an equation one has to find all of its solutions. Cramer's Rule is the straightforward way to find solutions by determinants of algebraic equations. How to find solutions of linear equations by addition, subtraction or substitution is explained. The formula solving quadratic equations is derived.

An *exponent* n is a symbol written above, and on the right of another symbol known as the *base* x as in x^n. All arithmetic operations apply to exponents.

The Binomial Theorem shows how to expand $(a+x)^n$ when a and n are any numbers, positive, negative, integral or fractional. When a=1 expansions of $(1+x)^n$ provide useful approximations.

The Exponential b^x and Logarithmic Functions $\log_b x$ were created to solve problems not solvable by known functions. The functions are used in every branch of mathematics. From a numerical point of view logarithms may be the most important arithmetic concept in mathematics.

Many problems are simplified when a rational function, the ratio of two polynomials, is decomposed into a sum of partial fractions with denominators of lower degree. Partial fractions have many applications such as simplifying many algebraic problems as well as the inverse Laplace Transform process.

The mathematical induction method of *proof by induction* has many uses such as proving theorems, discovering new results, and providing relatively simple proofs of theorems obtained by other means.

Arithmetic and geometric progressions are discussed.

Matrix algebra allows one to write and process equations efficiently. Furthermore, in many problems, the matrix format makes the next step easier to perceive.

A competent, serious, professional will tell you that acquiring knowledge is hard work. Specifically, many mathematicians will tell you that, even for them, learning and understanding mathematics at any level is not easy! Surprised?

Letters in Algebra In Arithmetic only numbers are used, because the goal is to learn how to manipulate numbers. Algebra is different. The goal is learning how to manipulate general equations.

The letters in the equations represent numbers. The letters represent words. We could write equations that are difficult-to-read using terms such as “first unknown”, “second unknown”, and so forth. Instead one letter words are used, words such as x, y, and z as in the equation $(x+y)(x-y)=x^2-y^2$.

Please remember this When working with algebraic expressions, one is manipulating *numbers*. The rules for manipulating algebraic expressions are consistent with the properties of the *number system*. Therefore, when manipulating the symbolic quantities of algebra always ask yourself "If I replace the symbols by numbers in my results are the results valid?" Many times the answer is not obvious.

The text In this modest mathematics text, we have tried hard to write in plain English. We do not use the phrase *this is obvious* for a good reason. Nothing is obvious to a person learning any subject.

Most Algebra books are forests full of trees that make it very difficult for the reader to know what is important in a forest. This text contains very few trees.

Trigonometry
Trigonometric functions are periodic. This is why they are immensely important. Many physical phenomena are periodic such as sound waves, radio waves, alternating electric current and voltage.

Originally, the primary object of the mathematics of plane trigonometry was a study of the relationships between sides and angles of a triangle, which evolved into definitions of trigonometric functions and a method of solving plane triangles. Today solving plane triangles only plays a minor role in modern engineering and scientific activities.

A plane triangle has three sides and three angles in a plane. If the values of any three of these six parts is a given, then solving the triangle means finding the values of the other three parts. In order to solve triangles the functions of angular magnitude, sine and cosine, were introduced. Later Plane Trigonometry was extended to include investigation of the properties of functions in addition to the sine and cosine, such as tangent, cotangent, secant, cosecant, and their inverses.

Angles are defined, and the degree and radian units are explained.

Pythagoras' Theorem is explained and the Pythagorean Identities are derived.

Solving Right Triangles is a straightforward process – 4 cases.

Algebra

Triangle Laws allow for straightforward solutions of oblique triangles.

In order to work with the trigonometric functions of any angle the trigonometric sine and cosine functions are defined by using a circle of radius 1.

The Inverse Trigonometric Functions are defined. The angle of a trigonometric function is the inverse of that function.

Equations for circular functions of two or more angles are derived.

The derivatives of the sine and cosine are derived.

The integrals of the sine and cosine are derived.

The cosh and sinh hyperbolic functions are derived.

There are a very large number of trigonometric identities. Do not memorize an equation. If you memorize a formula by a rote process, you may not remember it correctly. The long view is more effective. I.e. *understand* how the trig functions work. Then in effect you will "memorize" basic equations, and will be able to derive, in a very simple way, whatever you will need using Euler's identity (Trig Section 8.3).

Calculators Once upon a time we had to know how to use the sine-cosine table, the tangent table, the logarithm table, the log-sin-cos table, and the log-tan table. Today we use calculators!

Important: do NOT memorize!
Derive from basic definitions!

Contents

Trigonometry

1 Fundamental Operations

Real Number System: The world made a very wise decision when it decided not to introduce new symbols beyond the symbols 0 to 9. The world did not want to repeat the disaster of the Roman number system (such as XLVII or 47).

Knowing how to count is a prerequisite to any mathematical activity. Counting is ground zero. If you select any number, *the next number is found by adding 1*. In the abstract language of algebra, if letter n is shorthand for the word number and n represents any number, then the next number is $n+1$. This means there is no limit to the magnitude of a number.

Infinity is an alias for *a number that is as large as you please.*

The necessity of counting led to the introduction of the *positive integers*, the natural numbers, such as 1, 2, 3, 4, 5, … . This was followed by the introduction of the *positive rational numbers* such as ½, 5/9, 27/45. A rational number is defined as the ratio of two integers. Euclid's geometry showed the existence of irrational numbers such as $\sqrt{2}$. These numbers together with the *negative rational numbers* and the *negative irrational numbers* form the real number system.

Letters and other symbols are used to represent numbers. Thus in the formula for area $A = b \times h$ the letters b and h represent the number of units in the two sides of the area's rectangle. Hence b and h are implicit numbers that are called literal or general numbers, which are distinct from explicit numbers such as 4, 22, 7/8, 34.98.

In what follows literal numbers are assumed to belong to the real number system.

A Number Line: The very important *number line* is a graphic display of real numbers.

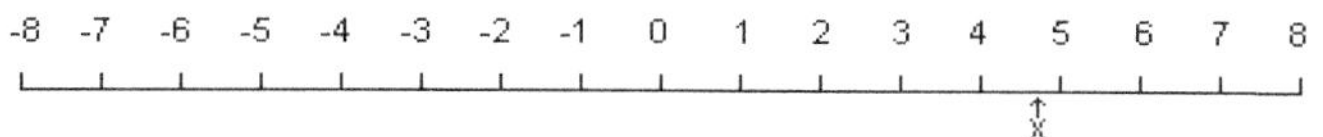

The number line is constructed by marking equal lengths along the line. Each mark on the number line is assigned a number. Assign 0 to any mark. Next, assign 1 to the first mark to the right of zero. Then the *distance* from 0 to 1 *represents* 1 unit of length. Subsequent marks 2, 3, 4, etc to the right add 1 unit to the distance. In the same way mark equal lengths to the left of zero, and label them –1, –2, –3, etc. Here x points to about 4.7 on the number line.

Now one can say, without proof, that to every explicit real number x there corresponds one and only one point on the number line. And, conversely to every point on the number line there corresponds one and only one explicit real number.

The Real Number System

The real number system is a composite of the following subsets.

N The natural numbers
{1, 2, 3, 4, 5, ….}
This is the set of numbers used for counting (a.k.a counting numbers).

W The whole numbers
{0, 1, 2, 3, 4, 5, ….}
This set adds zero to the set of natural numbers.

Z The integers.
{…., –6, –5, –4, –3, –2, –1, 0, 1, 2, 3, 4, 5, ….}
This set adds the negative integers to the set of whole numbers.

Q The rational numbers.
The set of rational numbers is the set of all numbers expressed as the ratio of two integers p/q where $q \neq 0$.
For example 1/3, 5/7, –213/5899, –2, (–2/1), etc

I The irrational numbers.
The set of all numbers that have decimal representations that do not terminate nor repeat. This is the set of all numbers that is *not* expressed as the ratio of two integers p/q where $q \neq 0$.

Fundamental Operations: The fundamental operations of algebra are addition, multiplication, subtraction, division, involution (raising to powers), evolution (extracting of roots). These operations are the same operations of arithmetic extended to the real number system.

As in arithmetic, division by zero is excluded. And application of evolution to negative numbers creates imaginary numbers.

Fundamental Assumptions of Algebra: Felix Klein[1] enumerates the eleven laws all elementary reckoning can be based on. There are *five fundamental laws* upon which *addition* depends. For any numbers x, y, z

(1) $x+y$ *is always a number* (*addition is always possible*)

(2) $x+y$ *is one valued*

(3) *The associative law holds* $(x+y)+z=x+(y+z)$

(4) *The commutative law holds* $x+y=y+x$

(5) *The monotonic law holds* *If* $y>z$, *then* $x+y>x+z$

There are *five exactly analogous laws* upon which *multiplication* depends.

(6) $x\times y$ *is always a number* (*multiplication is always possible*)

(7) $x\times y$ *is one valued*

(8) *The associative law holds* $(x\times y)\times z=x\times(y\times z)$

(9) *The commutative law holds* $x\times y=y\times x$

(10) *The monotonic law holds* *If* $y>z$, *then* $x\times y>x\times z$

Multiplication is connected to addition by the distributive law.

(11) $x\times(y+z)=x\times y+x\times z$

Klein claims "that it is easy to show that all elementary reckoning can be based on these eleven laws." We take his word for it, because proving this statement requires a very big digression. Klein does refer to the content of original sources, which one can pursue. There are many reasons for knowing and understanding the eleven laws. Here are two.

1. We use them all of the time without our being explicitly aware we are doing that, because their application is not explicit.

2. These laws play a central role in mathematics. They are hidden behind the scene, so to speak, of all operations as they are executed.

---

[1] Felix Klein 1908, "Arithmetic, Algebra, Analysis", ISBN 048643480X

Algebraic Expressions: Any symbol or combination of symbols that represents one or more numbers is referred to as an algebraic expression,

If an expression consists of several parts that are connected by plus and minus signs, than each of these parts together with the sign preceding each part is referred to as a *term*.

If two or more terms are multiplied together, then a *product* is formed and each term in the product is a *factor* of the product. Numbers w, x, y, z are factors of the product wxyz.

Numbers such as these are *composite* numbers.

$6 = 3 \times 2$

$12 = 2 \times 2 \times 3$

$150 = 5 \times 2 \times 5 \times 3$

Multiplication by 1 does not change a number. This is why 1 is referred to as a trivial factor.

Observe that there are numbers which do not have factors except the trivial factor 1. They are *prime* numbers.

$2, 3, 5, 7, 11, 13, 17, 19, 23, 29, 31, 37, \ .... \ are\ primes$

All *composite numbers are products of primes.* This follows from the Fundamental Theorem of Arithmetic, which has been proven to be true.

Powers and Exponents: An *exponent x* is a symbol written above, and on the right of, another symbol known as the *base b* as in b^x. The expression b^x is referred to as a power; specifically the xth power of b. All arithmetic operations apply to exponents. *The exponent can be any type of number.*

Many famous physical constants are usually expressed as a number in the range 1 to 9.999..... times a power of 10.

$$
\begin{array}{ll}
velocity\ of\ light & c = 2.997925 \times 10^{8}\ meters/second \\
Avogadro's\ number & N_A = 6.0225 \times 10^{23}\ mole^{-1} \\
charge\ of\ the\ electron & e = 1.60210 \times 10^{-19}\ Coulomb \\
Planck's\ constant & h = 6.62517 \times 10^{-34}\ Joule \cdot second \\
Boltzman's\ constant & k = 1.3805 \times 10^{-23}\ Joule/degree
\end{array}
$$

When the exponent is a positive integer, it indicates the number of times the base is a factor. For example:

$b^5 = b \times b \times b \times b \times b \quad (-a)^3 = (-a)(-a)(-a) = (-1)^3(a)^3 = -a^3$

Grouping Symbols: When the numbers in an algebraic expression are grouped correctly, the expression is simplified. Typical grouping symbols are parenthesis ( ), brackets [ ], and braces { }. Any symbols can be used.

Negative signs need special care. For example:

$$\begin{aligned} f &= -\{2x - [2(y-3x) - (x-5y)]\} \\ &= -\{2x - [2y - 6x - x + 5y]\} \\ &= -\{2x - [7y - 7x]\} \\ &= -\{2x - 7y + 7x\} \\ &= -9x + 7y \end{aligned}$$

Multiplication: Multiplication is implemented in three levels.

Level 0 term times term
Level 1 term times polynomial
Level 3 polynomial times polynomial

The last step adds the polynomials created during level 1. For example;

$\text{Level 0} \quad 2a^2 \times ab = 2a^3b$

$\text{Level 1} \quad 2a^2 \times (a^2 - 2ab - b^2) = 2a^4 - 4a^3b - 2a^2b^2$

$$\begin{array}{lrl} \text{Level 2} & a^2 \quad -2ab \quad -b^2 & \text{(multiplicand)} \\ & \underline{2a^2 \quad -ab \quad +b^2} & \text{(multiplier)} \\ & a^2b^2 - 2ab^3 - b^4 & \\ & -a^3b + 2a^2b^2 + ab^3 \qquad & \\ & \underline{2a^4 - 4a^3b - 2a^2b^2 \qquad\qquad} & \\ & 2a^4 - 5a^3b \ + a^2b^2 - ab^3 - b^4 & \text{(product)} \end{array}$$

$$\begin{array}{r} 5a^3 + 8a^2 - 23a - 1 \\ \underline{5a^2 - 7a - 2} \\ -10a^3 - 16a^2 + 46a + 2 \\ -35a^4 - 56a^3 + 161a^2 + 7a \qquad \\ \underline{25a^5 + 40a^4 - 115a^3 - 5a^2 \qquad\qquad} \\ 25a^5 + 05a^4 - 181a^3 + 140a^2 + 53a + 2 \end{array}$$

Division: Divide a polynomial by a polynomial by dividing the first term of the dividend by the first term of the divisor to produce the first term of the quotient.

Multiply the first term of the quotient by the divisor. Subtract the product from the dividend to obtain a new dividend. For example:

$$
\begin{array}{rl}
 & a \quad +3 \quad \text{(quotient)} \\
\text{(divisor)}\ 5a^2 - 7a - 2 & \overline{\big)\,5a^3 + 8a^2 - 23a - 1} \quad \text{(dividend)} \\
 & \underline{5a^3 - 7a^2 - 2a} \\
 & \quad 15a^2 - 21a - 1 \\
 & \quad \underline{15a^2 - 21a - 6} \\
 & \qquad\qquad 5 \quad \text{(remainder)}
\end{array}
$$

Divide y^3+27x^3 by $y+3x$. Note how the zeros extend y^3+27x^3.

$$
\begin{array}{rl}
 & y^2 - 3xy \quad + 9x^2 \\
\text{(divisor)}\ y + 3x & \overline{\big)\,y^3 + 0xy^2 + 0x^2y + 27x^3} \quad \text{(dividend)} \\
 & \underline{y^3 + 3xy^2} \\
 & \quad -3xy^2 + 0x^2y \\
 & \quad \underline{-3xy^2 - 9x^2y} \\
 & \qquad\quad 9x^2y + 27x^3 \\
 & \qquad\quad \underline{9x^2y + 27x^3} \\
 & \qquad\qquad\quad 0 \quad \text{(remainder)}
\end{array}
$$

$$
\begin{array}{rl}
 & x^2 - x + 1 \\
\text{(divisor)}\ x^4 + x^3 + 0 + 0 + 1 & \overline{\big)\,x^6 + 0 \quad + 0 + 0 + x^2 + x} \quad \text{(dividend)} \\
 & \underline{x^6 + x^5 + 0 + 0 + x^2} \\
 & \quad -x^5 \qquad\qquad + x \\
 & \quad \underline{-x^5 - x^4 \qquad - x} \\
 & \qquad\quad x^4 \qquad\quad + 2x \\
 & \qquad\quad \underline{x^4 + x^3 \qquad\quad + 1} \\
 & \qquad\qquad -x^3 \quad + 2x - 1 \quad \text{(remainder)}
\end{array}
$$

2 Equations

Algebra is a branch of mathematics which processes the relations and properties of numbers by means of letters, signs of operations, and other symbols. Algebra is probably the reader's introduction to abstraction where the general, such as the letters *x, y*, are in use as opposed to the particular such as specific numbers *n= 5 or 12.34 or -7 or -1008000.34.* Here is the guiding principle of algebraic operations.

The equality is not upset if the *same* change is made to both sides of =.

Solving Equations: For example to solve for x in linear equation 1 one needs to isolate x on one side of =. If one subtracts 7 from both sides of equation 1, then x is isolated.

$$\text{(1)}\ \ x+7=9 \quad \rightarrow \quad x+7-7=9-7$$
$$\text{subtract 7 from both sides of } = \text{to get } x=2$$

Again, to solve for x in equation 2 one needs to isolate x on one side of =. If one adds 4 to both sides of equation 2, then x is isolated.

$$\text{(2)}\ \ x-4=8 \quad \rightarrow \quad x-4+4=8+4 \quad \text{add 4 to both sides of } = \text{ to get } x=12$$

The *sign* of a term is changed when it is moved to the other side of =. In equation 1 note how +7 became –7. In equation 2, –4 became +4.

Addition and multiplication are the fundamental operations. Subtraction and division are the inverse operations of addition and multiplication respectively. Multiplication is connected to addition by the distributive law $x\times(y+z)=x\times y+x\times z$. The distributive law *expands* expressions.

Solving for x in equation 3 requires several steps. Multiply both sides of = by 4. Apply the distributive law. Then add 12 to both sides of =.

$$\text{(3)}\ \ \frac{x}{4}-3=5 \quad \rightarrow \quad 4\times\left(\frac{x}{4}-3\right)=4\times5 \quad \textit{multiply both sides of } = \textit{ by } 4$$

$$\textit{to get } 4\times\frac{x}{4}-4\times3=4\times5 \quad \rightarrow \quad x-12=20 \quad \rightarrow \quad x=32$$

$$\textit{check } \frac{32}{4}-3=8-3=5 \ \ \textit{qed}$$

This time divide both sides of = by 7 to solve for x.

(4) $7x-3=5 \quad \rightarrow \quad \frac{1}{7}\times(7x-3)=\frac{1}{7}\times 5 \quad$ *divide both sides of* = *by* 7

to get $\frac{1}{7}\times 7x-\frac{1}{7}\times 3=\frac{1}{7}\times 5 \quad \rightarrow \quad x-\frac{3}{7}=\frac{5}{7} \quad \rightarrow \quad x=\frac{8}{7}$

check $7\frac{8}{7}-3=8-3=5$ *qed*

The next two examples create identities valid for all x. The process expands (multiplying out) to produce a sum of terms. All terms have coefficient 1 or –1. Making coefficients 1 explicit is always a useful tactic.

(5) $(x-y)(-x+y)$

$=x(-x+y)-y(-x+y)$ *apply distributive law*

$=-x^2+xy+yx-y^2$ *apply distributive law again*

$=-x^2+2xy-y^2$ *add equal terms* $xy=yx$

$=(-1)x^2+(1)2xy+(-1)y^2$ *make* 1 *coefficients explicit*

$=(-1)x^2+(-1)(-1)2xy+(-1)y^2 \quad 1=(-1)(-1)$

$=(-1)[x^2+(-1)2xy+y^2]$ *factor out* -1, *a reverse distribution*

$=-[x^2-2xy+y^2] \quad -=(-1)$

(6) $(x-y)^2=(x-y)(x-y)$

$=x(x-y)-y(x-y)$ *apply distributive law*

$=x^2-xy-yx+y^2$ *apply distributive law again*

$=x^2-2xy+y^2$ *add equal terms*

Here is a closer look at part of equation 6.

(7) $-y(x-y)=(-y)\times(x)+(-y)\times(-y)$ *apply distributive law*

$-y(x-y)=(-1)yx+(-1)(-1)yy$ *make coefficients* 1 *explicit*

$-y(x-y)=-yx+y^2$

common error $\quad -y(x-y)=-yx-y^2$

Problem 201 Show that $x^2-y^2=(x-y)(x+y)$

Problem 202 Show that $x^3-y^3=(x-y)(x^2+xy+y^2)$

Word problems are used as means to learn how to translate text into equations. Consider this one.

A number x is tripled. Then 11 is added so that the sum equals 32. Solve for x.

$$(8)\ \ 3x+11=32 \rightarrow 3x+11-11=32-11 \rightarrow 3x=21 \rightarrow \frac{1}{3}3x=\frac{1}{3}21 \rightarrow x=7$$

Algebraic Operations on an Equation Analysis of a problem produces an equation. Usually the equation needs to be recast into some other form. Recasting is done by algebraic operations such as addition, multiplication, subtraction, and division. Three examples follow.

(9) Newton's distance equation : $s = vt + \frac{1}{2}at^2$ solve for v

subtract $\frac{1}{2}at^2$ from both sides $s-\frac{1}{2}at^2 = vt+\frac{1}{2}at^2-\frac{1}{2}at^2 \rightarrow s-\frac{1}{2}at^2 = vt$

divide both sides by t $\frac{s}{t}-\frac{at^2}{2t}=\frac{vt}{t} \rightarrow v=\frac{s}{t}-\frac{at}{2}$

(10) $\sin^2 x+\cos^2 x=1$ solve for cos x

subtract $\sin^2 x$ from both sides : $\sin^2 x+\cos^2 x-\sin^2 x=1-\sin^2 x$

$\cos^2 x=1-\sin^2 x$

$\cos x=\left(1-\sin^2 x\right)^{\frac{1}{2}}$

(11) $\cos\frac{x}{2}=\left(\frac{1+\cos x}{2}\right)^{\frac{1}{2}}$ solve for cos x

square both sides $\cos^2\frac{x}{2}=\frac{1+\cos x}{2}=\frac{1}{2}+\frac{\cos x}{2}$

subtract $\frac{1}{2}$ from both sides $\cos^2\frac{x}{2}-\frac{1}{2}=\frac{1}{2}+\frac{\cos x}{2}-\frac{1}{2} \rightarrow \cos^2\frac{x}{2}-\frac{1}{2}=\frac{\cos x}{2}$

multiply both sides by 2 $2\cos^2\frac{x}{2}-2\frac{1}{2}=2\frac{\cos x}{2} \rightarrow 2\cos^2\frac{x}{2}-1=\cos x$

Cartesian coordinate system: Algebraic problems are simplified and clarified by interpreting expressions and equations geometrically by representing lines, curves, and surfaces in a coordinate system.

Rene Descartes, a French mathematician, published in 1637 the first systematic work on merging the concepts of algebra and a geometric coordinate system now known as the Cartesian coordinate system.

The number line was introduced in the study of Arithmetic (Figure 201).

Figure 201 Part of a Number Line

-8 -7 -6 -5 -4 -3 -2 -1 0 1 2 3 4 5 6 7 8

The number line is constructed by marking off equal lengths along the line. Each mark on the number line is assigned a number. Assign 0 to any mark. Next, assign 1 to the first mark to the right of zero. Then the *distance* from 0 to 1 *represents* 1 unit of length. Subsequent marks to the right add 1 unit to the distance. Label subsequent marks 2, 3, 4, and so forth. Repeat for negative numbers to the left.

The *distance* from 0 to 1 *represents* 1. The distance from 1 to 2 represents 1 more. Watch this. The distance from –3 to –2 also represents 1 more. The distance from 2 to 1 represents 1 less. The distance from –1 to –2 also represents 1 less. Move to the right to increase value. Move to the left to decrease value.

If a second number line is rotated by 90 degrees and superimposed on the horizontal number line at coordinate (0,0) an x, y coordinate system is established (Figure 202).

An elementary example of algebra in action is the derivation of equation (9a) of *any* straight line where letters x and y represent variables (the coordinates of the line), and letters c and d represent constants.

$(9a)\quad y = cx + d \qquad (9b)\quad y = -3x + 7 \qquad (9c)\quad y = 5.2x - 3.9$

The key word here is *any*. Variables represent any numbers, whereas constants represent specific numbers that do not change. Equation 9a is *one* form of the general straight line equation. Equations 9b and 9c are two specific examples.

Figure 202 Cartesian Coordinate System

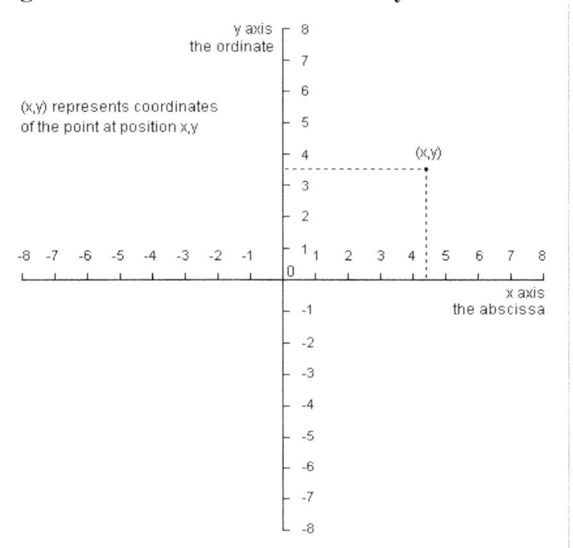

A line is drawn in the x, y coordinate system (Figure 203).
Four points are marked on the line.

(x,y)=(c,0)	x axis intercept
$(x,y)=(x_1,y_1)$	a specific point
(x,y)=(0,b)	y axis intercept
(x,y)=(x,y)	any point x, y in the (x,y) plane

Figure 203 Line in a Cartesian Coordinate System

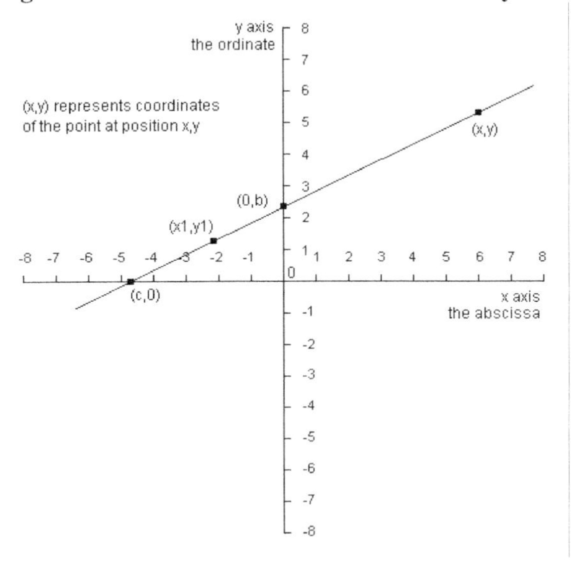

Applying Geometry: A move from point (x, y) to point (x_1, y_1) causes changes in x and y, which are described as increments Δx *and* Δy (delta x and delta y). *Δ is the universal mathematics symbol for change in value.*

The slope *m* of a line is a measure of how steep a line is. A horizontal line has slope equal to zero ($\Delta y=0$). A slope m=1 means every step in x is matched by an equal step in y ($\Delta y = \Delta x$). The equation for slope m is defined as the ratio $\Delta y/\Delta x$.

$$(12)\quad m = \frac{\text{increment in y}}{\text{increment in x}} = \frac{\Delta y}{\Delta x} = \frac{y - y_1}{x - x_1} \quad \text{created by points } (x, y) \text{ and } (x_1, y_1)$$

As in arithmetic the four fundamental operations of algebra are addition, multiplication, subtraction and division.

An equation equates expression A to expression B so that A=B.

Algebraic operations produce the equation of the line (Figure 203). Multiply both sides of = in equation 10 by ($x-x_1$).

$$(13)\quad (x - x_1)m = \frac{y - y_1}{x - x_1}(x - x_1) \;\rightarrow\; (x - x_1)m = (y - y_1)$$

Add y_1 to both sides of =.

$$(14)\quad (x - x_1)m + y_1 = y - y_1 + y_1 \quad \rightarrow \quad (x - x_1)m + y_1 = y$$

$$(15)\quad y = m(x - x_1) + y_1$$

Substitute point (0, b) for point (x_1, y_1) in equation 13. I.e. substitute 0 for x_1 and b for y_1.

$$(16)\quad y = mx + b \quad \text{the slope intercept form of a line equation}$$

Calculate the value of slope m. In equation 14 substitute point (c,0) for point (x,y). I.e. substitute c for x and 0 for y.

(17a) $0 = mc + b$

(17b) subtract b from both sides of = $\rightarrow$ $0 - b = mc + b - b$ $\rightarrow$ $-b = mc$

(17c) divide both sides of = by c $\rightarrow$ $m = -b\frac{1}{c} = -\frac{b}{c}$

Note: c is a negative number, which makes m a positive number

As an alternative *define* slope m by using x axis and y axis intercept points (c,0) and (0,b).

(18) $m = \frac{\Delta y}{\Delta x} = \frac{b-0}{0-c} = \frac{b}{-c}$ positive slope of the line

Yet another way is to let any point (x,y) and x axis intercept point (c,0) define the line equation's intercept form (equation 17i).

(19a) $m = \frac{b}{-c} = \frac{\Delta y}{\Delta x} = \frac{y-0}{x-c}$

(19b) multiply both sides of = by $x - c$ $\rightarrow$ $\frac{b}{-c}(x-c) = \frac{y-0}{x-c}(x-c)$

(19c) cancel $(x-c)/(x-c) = 1$, and multiply by $-c$ $\rightarrow$ $\frac{b}{-c}(x-c)(-c) = y(-c)$

(19d) cancel $(-c)/(-c) = 1$, distribute b $\rightarrow$ $bx - bc = -yc$

(19e) subtract bx from both sides of = $\rightarrow$ $bx - bc - bx = -yc - bx$

(19f) so that $-bc = -yc - bx$

(19g) multiply both sides of = by -1 $\rightarrow$ $bc = yc + bx$

(19h) divide both sides of = by bc $\rightarrow$ $\frac{bc}{bc} = \frac{yc}{bc} + \frac{bx}{bc}$

(19i) cancel $\frac{bc}{bc} = 1$, $\frac{c}{c} = 1$, $\frac{b}{b} = 1$ $\rightarrow$ $1 = \frac{y}{b} + \frac{x}{c}$ intercept form

Looking at Figure 203, clearly the slope of the line m is positive. Slope m is positive, because numbers b and –c, repeat –c, are both positive.

There was no problem using (any number) c while developing equations. The fact that c is negative did not matter, which emphasizes what algebra is all about.

Temperature Scales

One application of the straight line is the relationship of the Fahrenheit temperature scale T_F to the Centigrade temperature scale T_C.

Water freezes at 32° and boils at 212° on the Fahrenheit temperature scale T_F, whereas it freezes at 0° and boils at 100° on the Centigrade temperature scale T_C. The slope of the line is the ratio of the two temperature increments 212–32 and 100–0. Designate T_F as the y axis and T_C as the x axis.

$$(1)\quad m = \frac{increment\ in\ T_F}{increment\ in\ T_C} = \frac{212-32}{100-0} = \frac{180}{100} = \frac{9}{5}$$

Any point (T_F,T_C) and T_F axis intercept point (32,0) define the temperature equation's slope.

$$(2a)\quad m = \frac{\Delta T_F}{\Delta T_C} \quad \rightarrow \quad \frac{9}{5} = \frac{T_F - 32}{T_C - 0}$$

$$(2b)\quad multiply\ both\ sides\ of\ =\ by\ T_C - 0 \quad \rightarrow \quad \frac{9}{5}(T_C - 0) = \frac{T_F - 32}{T_C - 0}(T_C - 0)$$

$$(2c)\quad cancel\ (T_C - 0) \quad \rightarrow \quad \frac{9}{5}(T_C - 0) = T_F - 32$$

$$(2d)\quad add\ 32\ to\ both\ sides\ of\ = \quad \rightarrow \quad \frac{9}{5}(T_C - 0) + 32 = T_F - 32 + 32$$

$$(2e)\quad cancel\ 32,\ distribute\ \frac{9}{5} \quad \rightarrow \quad T_F = \frac{9}{5}T_C + 32$$

Problem 203 Show that $x^3 + y^3 = (x+y)(x^2 - xy + y^2)$

Problem 204 Show that $x^3 + y^3 = x^3 + 3x^2y + 3xy^2 + y^3$

Problem 205 Show that $(x+y+z)^2 = x^2 + y^2 + z^2 + 2yz + 22zx + 2xy$

Problem 206 Show that

$$if\ \frac{a}{b} = \frac{c}{d}\ and\ k = \frac{a}{b} = \frac{c}{d}\ and\ k \neq 0,\ then\ \frac{a+b}{a-b} = \frac{c+d}{c-d}$$

Problem 207 Plot points (q, 2q) where q=0, ±1, ±2, ±3, ±4

Electrical Feedback Amplifier

Analysis of an electrical feedback amplifier (Figure 801) produces three equations 1.

Figure 801 Feedback Amplifier

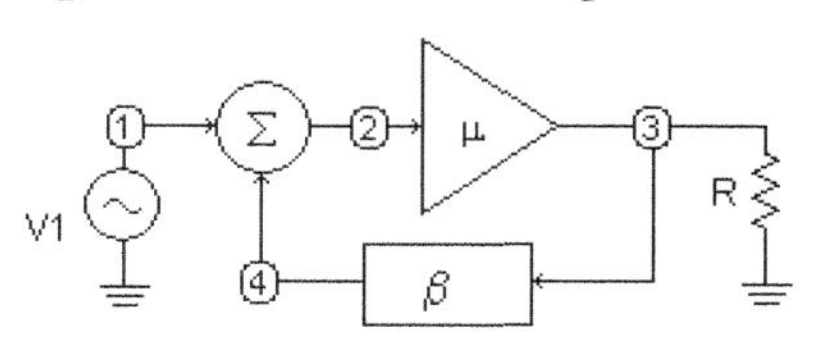

(1*a*) $v_3 = \mu v_2$

(1*b*) $v_2 = v_1 + v_4$

(1*c*) $v_4 = \beta v_3$

What follows is a typical sequence of many different operations such as replacement, distribution, subtraction, factoring, and division.

Solve for v_3/v_1. First v_2 is eliminated by substitution (replacement). Second v_4 is eliminated producing an equation in v_3 and v_1. Then solve for v_3/v_1.

(2*a*) $v_3 = \mu v_2$

$v_3 = \mu(v_1 + v_4)$ — *replace* v_2 *with* $v_1 + v_4$ (1*b*)

$v_3 = \mu(v_1 + \beta v_3)$ — *replace* v_4 *with* βv_3 (1*c*)

$v_3 = \mu v_1 + \mu\beta v_3$ — *use distributive law*

(2*b*) $v_3 - \mu\beta v_3 = \mu v_1 + \mu\beta v_3 - \mu\beta v_3$ — *subtract* $\mu\beta v_3$ *from both sides of* =

$v_3 - \mu\beta v_3 = \mu v_1$

$v_3(1 - \mu\beta) = \mu v_1$ — *factor out* v_3

(2*c*) $v_3 = \dfrac{\mu v_1}{1 - \mu\beta}$ — *divide both sides of* = *by* $1 - \mu\beta$

$\dfrac{v_3}{v_1} = \dfrac{\mu}{1 - \mu\beta}$ — *divide both sides of* = *by* v_1

Problem 208 Plot points (x, y) where x= 0, ±1, ±2 and $y=x^2+1$.
Problem 209 Plot points (x, y) where x ≤ 8, and y = –2.
Problem 210 Plot points (x, y) where x = 3, and y = –2 to y = +5.
Problem 211 Plot points (q, 2q) where q=0, ±1, ±2, ±3, ±4
Problem 212 Show that the equation for points (x, y) = (–4, –1), (–2, 1), (0, 3), (2, 5) is y = x+3.

3 Polynomials

If x is a *variable*, then x can be *any* number real or complex. If a_j is a *constant*, then a_j can be *any specific* number real or complex. One can build a *polynomial* as follows from degree 0 to degree n, where the *degree* equals the highest power of x.

(1*a*) $f_0(x) = a_0$

(1*b*) $f_1(x) = a_1 x + a_0$

(1*c*) $f_2(x) = a_2 x^2 + a_1 x + a_0$

and so forth until the mathematician's favorite degree n is reached

(1*n*) $f_n(x) = a_n x^n + a_{n-1} x^{n-1} + \cdots + a_2 x^2 + a_1 x + a_0$

Reminder: what is a power of x such as x^n when n is an integer?

(2) $x^n = x \times x \times x \times \cdots \times x \times x$ *(the product of n x's)*

Next build a polynomial in a way that emphasizes *roots*. A root r of a polynomial is a value of x that makes the polynomial equal zero.

(3a) $f_1(x) = (x - r_1)$

(3b) $f_2(x) = (x - r_1)(x - r_2)$

and so forth until the mathematician's favorite degree n is reached

(3n) $f_n(x) = (x - r_1)(x - r_2) \cdots (x - r_{n-1})(x - r_n)$

Apply the distributive law to equation 3b.

(4)
$$\begin{aligned} f_2(x) &= (x - r_1)(x - r_2) \\ &= x(x - r_2) - r_1(x - r_2) \\ &= x^2 - r_2 x - r_1 x + r_1 r_2 \\ &= x^2 - (r_2 + r_1)x + r_1 r_2 \end{aligned}$$

Equate equations 1c and 4 in order to relate roots to coefficients.

(5*a*) 1*c* $f_2(x) = 4\ f_2(x)$

(5*b*) $a_2 x^2 + a_1 x + a_0 = x^2 - (r_2 + r_1)x + r_1 r_2$

(5*c*) $(a_2 - 1)x^2 + (a_1 + (r_2 + r_1))x + (a_0 - r_1 r_2) = 0$

(5*d*) *so that for any x* $\quad a_2 = 1, \quad a_1 = -(r_2 + r_1), \quad a_0 = r_1 r_2$

3.1 Operations

Polynomial addition and subtraction Addition is implemented by grouping of powers of x (equation 6b), and adding the grouped coefficients (equation 6d). Subtraction is implemented by grouping of powers of x and subtracting the coefficients. Another way to add polynomials is to use arithmetic format and add each column (equation 6e). There are no carries.

add the polynomials

(6a) $(2x^5 - 7x^4 + x^3 + 3x^2 - 2x + 1) + (-x^5 + 3x^4 - 16x^2 - 4x + 5)$

group by powers of x

(6b) $(2x^5 - x^5) + (-7x^4 + 3x^4) + (x^3) + (3x^2 - 16x^2) + (-2x - 4x) + (1 + 5)$

apply the distributive law in reverse, factor out the powers of x

(6c) $(2-1)x^5 + (-7+3)x^4 + (1+0)x^3 + (3-16)x^2 + (-2-4)x + (1+5)$

add the coefficients

(6d) $x^5 - 4x^4 + x^3 - 13x^2 - 6x + 6$

AN EASIER WAY

redo the addition in Arithmetic format, add term by term

$$\begin{array}{lr} & 2x^5 - 7x^4 + x^3 + 3x^2 - 2x + 1 \\ \text{(6e)} & \underline{-x^5 + 3x^4 \qquad - 16x^2 - 4x + 5} \\ & x^5 - 4x^4 + x^3 - 13x^2 - 6x + 6 \end{array}$$

Polynomial multiplication in arithmetic format applies the distributive law in the same way as in arithmetic where each term now is a power of x instead of a number. The last step adds the rows of each column with no carries.

$$\begin{array}{lrr} \text{(7a)} & x^5 - x^3 + 2x + 1 & f(x) \\ \times & 2x^3 + 5x + 1 & g(x) \\ \hline & x^5 \quad - x^3 \quad + 2x + 1 & 1 \times f(x) \\ & 5x^6 \quad - 5x^4 \quad + 10x^2 + 5x \quad & 5x \times f(x) \\ & 2x^8 - 2x^6 \quad + 4x^4 + 2x^3 \qquad\qquad & 2x^3 \times f(x) \\ \hline & 2x^8 + 3x^6 + x^5 - x^4 + x^3 + 10x^2 + 7x + 1 & \text{sum of 3 rows} \end{array}$$

(7b) *check – sum of 3 rows* $S = 1 \times f(x) + 5x \times f(x) + 2x^3 \times f(x)$

(7c) *factor out f(x)* $S = f(x)(1 + 5x + 2x^3) = f(x) \times g(x)$ *qed*

Multiplication of 2 polynomials that uses the distributive law directly.

Polynomials to multiply

(8*a*) $(2x^3+5x+1)(x^5-x^3+2x+1)$

apply the distributive law

(8*b*) $2x^3(x^5-x^3+2x+1)+(5x)(x^5-x^3+2x+1)+(1)(x^5-x^3+2x+1)$

multiply

(8*c*) $(2x^8-2x^6+4x^4+2x^3)+(5x^6-5x^4+10x^2+5x)+(x^5-x^3+2x+1)$

gather coefficients of powers of x

(8*d*) $(2)x^8+(-2+5)x^6+(1)x^5+(4-5)x^4+(2-1)x^3+(10)x^2+(5+2)x+1$

add coefficients

(8*e*) $2x^8+3x^6+x^5-x^4+x^3+10x^2+7x+1$

Polynomial Division Divide h(x) by p(x) to get quotient q(x) and remainder r(x). Use the polynomials in equations 8a and 8e. The key is that the quotient term x^5 multiplied by p(x) produces $2x^8$ as the first term, which subtracts from $2x^8$ in h(x). The value of the quotient terms x^5, $-x^3$, 0, 2x, 1 is determined by the requirement to cancel the next term in h(x).

$$(9)\quad \frac{h(x)}{p(x)}=q(x)+\frac{r(x)}{p(x)} \rightarrow \begin{array}{r} q(x) \\ p(x)\overline{\smash{)}h(x)} \end{array} \text{ where } r(x) \text{ appears in the last step}$$

$$(10)\quad \begin{array}{rl}
x^5 - x^3 + 0 + 2x + 1 & \textit{quotient} \\
2x^3+5x+1\overline{\smash{)}2x^8+3x^6+x^5-x^4+x^3+10x^2+7x+1} & h(x)/p(x) \\
\underline{2x^8+5x^6+x^5} \qquad\qquad\qquad\qquad & x^5p(x) \\
0-2x^6+0-x^4+x^3+10x^2+7x+1 & \textit{subtract} \\
\underline{-2x^6 \quad -5x^4-x^3} \qquad\qquad\quad & -x^3p(x) \\
0+0+4x^4+2x^3+10x^2+7x+1 & \textit{subtract} \\
\underline{4x^4 \qquad +10x^2+2x} & 2xp(x) \\
0+2x^3 \qquad +5x+1 & \textit{subtract} \\
\underline{2x^3 \qquad +5x+1} & 1\times p(x) \\
0 & \textit{subtract}
\end{array}$$

Problem 301 Divide $f(x)=x^3-3x^2+4$ *by* $x-2$ to get quotient x^2-x-2 and remainder 0.

Problem 302 Divide $f(x)=3x^3-13x^2+13x-3$ *by* $x-1$ to get quotient $3x^2-10x+3$ and remainder 0.

3.2 The Remainder Theorem

In arithmetic division of x by y produced a quotient q and a remainder r where $x = qy + r$. The quotient q equals the number of y's in x and the remainder r equals a partial y that is left over. For example there are 196 y's (23's) in 4521 and a partial y of 13 is left over.

$$(11a)\quad \frac{dividend}{divisor} = quotient + \frac{remainder}{divisor} \;\rightarrow\; \frac{x}{y} = q + \frac{r}{y} \;\rightarrow\; \frac{4521}{23} = 196 + \frac{13}{23}$$

$$(11b)\quad x = qy + r \quad \rightarrow \quad 4521 = 196 \times 23 + 13$$

One way to determine whether or not a number is a root of a polynomial is to use the division process.

For example to find out whether or not 2 is a root of polynomial h(x) divide h(x) by (x–2) until no x appears in the remainder. I.e. the remainder r(x) is a number. If the number is 0 then 2 is a root. Since the remainder is 32 two is not a root.

$$(12)\quad \begin{array}{r|l} & x^2 + 7x + 10 \\ \hline x-2 & x^3 + 5x^2 - 4x + 12 \\ & \underline{x^3 - 2x^2} \\ & \quad 7x^2 - 4x \\ & \quad \underline{7x^2 - 14x} \\ & \qquad 10x + 12 \\ & \qquad \underline{10x - 20} \\ & \qquad\qquad 32 \end{array}$$

An equation for h(x) that is valid for all x is

$$(13)\quad h(x) = p(x)q(x) + r(x)$$

$$h(x) = (x-2)(x^2 + 7x + 10) + 32 \quad and \quad h(2) = 32 = r(2)$$

Use the Remainder Theorem to avoid the division process in eqn 12.

Remainder Theorem 1 *The remainder r(b) when the polynomial* $f(x)=a_nx^n+a_{n-1}x^{n-1}+\cdots+a_2x^2+a_1x+a_0$ *is divided by (x–b) is* $r(b)=f(b)=a_nb^n+a_{n-1}b^{n-1}+\cdots+a_2b^2+a_1b+a_0$

Proof Let *q(x)* be the quotient, a polynomial of degree n–1 in *x*, and let *r(x)* be the remainder when *f(x)* is divided by *(x–b)*. Then

$$f(x)=(x-b)q(x)+r(x)$$

Since the equation is true for all values of x it is true when x=b, and so

$$f(b)=(b-b)q(b)+r(b) \quad \rightarrow \quad f(b)=r(b)$$

Remainder Theorem 2 *If f(b)=0 then the polynomial*

$$f(x)=a_nx^n+a_{n-1}x^{n-1}+\cdots+a_2x^2+a_1x+a_0$$

has (x–b) as a factor. Conversely if (x–b) is a factor of f(x), then f(b)=0.

Proof By Remainder Theorem 1 when *f(x)* is divided by *(x–b)* the remainder is *f(b)*.

$$f(b)=a_nb^n+a_{n-1}b^{n-1}+\cdots+a_2b^2+a_1b+a_0$$

If *f(b)=0* the remainder *r(b)=0*, then *(x–b)* divides *f(x)* exactly so that *(x–b)* is a *factor* of *f(x)*.

Conversely if *(x–b)* is a factor of *f(x)*, then *f(b)=0,* and the remainder *r(x)=r(b)* equals zero.

$$f(x)=(x-b)q(x)+r(x)$$
$$f(b)=(b-b)q(b)+r(b)=r(b) \quad \rightarrow \quad if\ f(b)=0,\ then\ r(b)=0$$

3.3 Factors of Polynomials and the Remainder Theorem

Factoring is the reverse of the distributive operation. Going from left to right in $(x-2)(x+3)=x^2+x-6$ is expanding. Going from right to left in $(x-2)(x+3)=x^2+x-6$ is factoring.

Find the roots (factors) of $h(x)=3x^3-13x^2+13x-3$. The constant term 3 equals the product of the roots (e.g. see a_0 in equation 5d on page 16). Therefore the possible roots of h(x) must include ±1, ±3, which are the factors of 3.

$$(14a)\quad h(1)=3-13+13-3=0$$
$$(14b)\quad h(-1)=-3-13-13-3\neq 0$$
$$(14c)\quad h(3)=3\cdot 27-13\cdot 9+13\cdot 3-3=81-117+39-3=0$$
$$(14d)\quad h(-3)=3\cdot(-27)-13\cdot 9+13\cdot(-3)-3=-81-117-39-3\neq 0$$
$$(14e)\quad h(x)=(x-1)(x-3)g(x)=(x^2-4x+3)g(x)$$

Find $g(x)=h(x)/(x^2-4x+3)$. One way is to divide.

$$(14f)\quad \begin{array}{r|l} & 3x-1 \\ \hline x^2-4x+3 & 3x^3-13x^2+13x-3 \\ & \underline{3x^3-12x^2+9x} \\ & \quad -x^2+4x-3 \\ & \quad \underline{-x^2+4x-3} \\ & \qquad\qquad 0 \end{array}$$

$$(14g)\quad g(x)=3x-1$$
$$(14h)\quad h(x)=(x-1)(x-3)(3x-1)$$

Factor restrictions: Polynomials are formed with integer coefficients and integral powers of x. Therefore the following factors are not allowed. The restrictions are not a problem, because factoring is stopped at (x–1) for example.

$$(15a)\quad (x-1)=\left(\sqrt{x}+1\right)\left(\sqrt{x}-1\right)$$
$$(15b)\quad (x-3)=\left(\sqrt{x}+\sqrt{3}\right)\left(\sqrt{x}-\sqrt{3}\right)$$

For example prove by means of the remainder theorem that (2x–1) is a factor of $f(x) = 2x^6 - 3x^5 + x^4 - 2x^2 + 3x - 1$

(16a) $(2x - 1) = 2\left(x - \frac{1}{2}\right)$ $\quad \frac{1}{2}$ is a root if $f(\frac{1}{2}) = 0$

(16b) $f(\frac{1}{2}) = 2\frac{1}{2}^6 - 3\frac{1}{2}^5 + \frac{1}{2}^4 - 2\frac{1}{2}^2 + 3\frac{1}{2} - 1 = \frac{1}{32} - \frac{3}{32} + \frac{1}{16} - \frac{1}{2} + \frac{3}{2} - 1 = 0$

Harder. Find the factors of $f(x, y) = 3x^3 - 22x^2 y + 43xy^2 - 12y^3$.

Think of this as a third degree polynomial in x whose coefficients of x are $3, -22y, 43y^2, -12y^3$.

Apply *Remainder Theorem 2.* Note that $-12y^3$ in f(x) is the product of the roots, which has the factors –12, ±4, ±3, y, y, y and combinations thereof.

(17*a*) $f(y, y) = 3y^3 - 22y^2 y + 43yy^2 - 12y^3 \neq 0$

(17*b*) $f(3y, y) = 3 \cdot 3^3 y^3 - 22 \cdot 3^2 y^3 + 43 \cdot 3y \cdot y^2 - 12y^3$

$= (81 - 198 + 129 - 12)y^3 = 0$ *and so* $(x - 3y)$ *is a factor*

Lowest Common Multiple (LCM): An expression that is a multiple of two or more expressions is referred to as a common multiple. The common multiple that has the smallest possible number of factors is the *lowest common multiple.* For example:

Find the LCM of $4x - 8, x^2 - 6x + 5, x^2 + 2x + 4$

$4x - 8 = 4(x - 2)$

$x^2 - 6x + 5 = (x - 1)(x - 5)$

$x^2 - 2x + 1 = (x - 1)(x - 1)$

$LCM = 4(x - 2)(x - 1)(x - 1)(x - 5)$

Problem 303 Divide 17a by (x–3y) to get $f(x, y) = (x - 3y)(3x^2 - 13yx + 4y^2)$.

Problem 304 Show that $f(x, y) = (x - 3y)(3x - y)(x - 4y)$

3.4 The Zeros of a Polynomial

We offer without proof the

Fundamental Theorem of Algebra Every integral rational equation in one variable x has at least one zero (root).

Corollary Every polynomial of degree n can be decomposed into n linear factors (roots).

Proof Let
$f(x) = a_n x^n + a_{n-1}x^{n-1} + \cdots + a_2 x^2 + a_1 x + a_0 \qquad (a_n \neq 0)$

By the *Fundamental Theorem* $f(x)=0$ has at least one zero z_1 so that $f(x) = (x - z_1)q_1(x)$.

By the *Fundamental Theorem* $q_1(x)=0$ has at least one zero z_2 so that $q_1(x) = (x - z_2)q_2(x)$ *and so* $f(x) = (x - z_1)(x - z_2)q_2(x)$.

Repeating the process n times
$f(x) = a_n(x - z_1)(x - z_2)\cdots(x - z_n)$ where the z_k are the n roots of f(x)

Theorem Every integral rational equation of degree n in one variable x has no more than n zeros (roots).

Proof Let m, n be any numbers, then from the *Corollary*
$f(x) = a_n(x - z_1)(x - z_2)\cdots(x - z_n)$

Clearly $f(x)$ cannot equal zero when x has any value m distinct from the n zeros z_k.

Note: The zeros need not all be real and distinct. For example if $z = z_3 = z_4 = z_9$, then z is a *multiple zero of order 3*. Nor do they have to be real. If a zero is a complex number $z=a+ib$, then some other zero has to be the complex conjugate $z=a-ib$ when the coefficients of $f(x)$ are real. In that case complex zeros always occur in conjugate pairs, and may be of any order. In any case the number of zeros cannot exceed n, the degree of $f(x)$.

3.5 Newton's Method for finding Zeros

Let $f(x)$ denote any polynomial in x and let $f'(x)$ denote its differential coefficient. Then

(18) $\dfrac{f(a+h)-f(a)}{h} \to f'(a)$ $\qquad$ $as\ h \to 0$

It follows that, when h is small

(19*a*) $f(a+h)-f(a) \approx hf'(a)$

(19*b*) $f(a+h) \approx f(a)+hf'(a)=0 \ \Rightarrow\ h=-\dfrac{f(a)}{f'(a)}$

We apply equation 19b to the problem of finding the numerical value of a zero of a polynomial ***If*** $\boldsymbol{f(x)=x^n}$***, then*** $\boldsymbol{f'(x)=nx^{n-1}}$.

(20*a*) *let* $f(x) \equiv x^3-3x^2+4x-3$

(20*b*) *so that* $f'(x)=3x^2-6x+4+0$

A table of values shows that there is a zero between x=1 and x=2, because f(x) changed sign from −1 to 1.

(21)

$x=$	0	1	2	3
$f(x)=$	−3	−1	1	9

Since the x values are 1 and 2 let the first estimate for the zero's value be a=1.5. Then from 19b

(22*a*) $f(1.5+h_1) \approx f(1.5)+h_1 f'(1.5) = -\frac{3}{8}+h_1\frac{7}{4}=0$

(22*b*) $h_l = \frac{3}{8}\frac{4}{7}=\frac{12}{56} \approx 0.2 \ \Rightarrow\ next\ estimate\ of\ zero = 1.7$

(22*c*) $f(1.7+h_2) \approx f(1.7)+h_2 f'(1.7) = 0.043+h_2 2.47=0$

(22*d*) $h_2 = -\frac{0.043}{2.47} \approx -0.02 \ \Rightarrow\ 3rd\ estimate\ of\ zero = 1.68\ and\ so\ forth$

Use Newton's method to find one *real* root r.

Problem 305 $x^7-5x+3=0$ (r=0.6060)

Problem 306 $x^5-3x^2-8=0$ (r=1.77075)

Problem 307 $x^3+7x-3=0$ (r=0.4180)

Problem 308 $70x^3-81x^2-100x+96=0$ (r=0.8000)

Problem 309 $14x^4+23x^3-16x^2+23x-30=0$ (0.860)

---

4 Polynomial Equations

An *equality* is a statement that two algebraic expressions are equal. The two expressions are referred to as members or sides of the equality. There are two kinds of equalities: identities and equations.

We have an *identical equality*, or simply an *identity*, if the two members of the equality are equal for *all values* of the symbols for which the members are defined. For example use the distributive law to show that equations 1a and 1b are identities.

$(1a)\ \ x^2 - y^2 = (x+y)(x-y)$ $\qquad$ $(1b)\ \ (2x-1)^2 + 4x = 4x^2 + 1$

We have an *equation*, if the two members of the equality are equal only for certain particular values of the symbols for which the members are defined. For example

$(2a)\ \ x - 3 = 2\ \ (only\ \ for\ \ x = 5)$ $\qquad$ $(2b)\ \ x^2 + 2 = 3x\ \ (only\ \ for\ \ x = 1\ and\ 2)$

An equation in one unknown x is an algebraic statement expressing a condition that the variable x must satisfy.

An equation in two unknowns x, y is an algebraic statement relating x and y that does not restrict either x *or* y to specific value(s). For example calculate y when x takes integer values 1, 2, 3, ...

$(3a)\ \ y = 5x + 3$ $\qquad$ $(3b)\ \ y = 5x^3 + 2x + 17$

Degree of a term Count the letters to find the degree of a term with respect to a specific letter. For example the term *xxyyyz* is a third degree term in y, second degree in x, and first degree in z. The degree of an equation with respect to any letter is the highest degree of that letter.

Solving an equation Solving an equation means finding *all* of its solutions. Any value of one unknown that makes both sides equal is a solution of the equation.

Allowable operations on an equation Adding the same number or expression to, or subtracting the same number or expression from, *both* sides. Multiplying or dividing *both* sides by the same number or expression, provided the divider is not zero. These steps lead to equivalent equations.

However, if both members of an equation are multiplied by an expression containing the unknown, or raised to the same integral power, extraneous roots not in the original equation may be introduced.

For example:

Multiply both sides of $x - 1 = 3$ by $(x + 4)$ to get $(x - 1)(x + 4) = 3(x + 4)$

Then $x^2 - 3x + 4 = 3x + 12$ has roots $+4$ and -4

Square both sides of $3x = 2x - 1$ to get $9x^2 = 4x^2 - 4x + 1$

or $5x^2 + 4x - 1 = 0$

Then the equation now has root 1/5 in addition to the original root -1

Solve $4 + \frac{x+3}{x-3} - \frac{4x^2}{x^2-9} = \frac{x-3}{x+3}$

Clear the fractions by using the lowest common denoinator $x^2 - 9$

$$4(x^2 - 9) + (x + 3)^2 - 4x^2 = (x - 3)^2$$

$$4x^2 - 36 + x^2 + 6x + 9 - 4x^2 = x^2 - 6x + 9$$

$$-36 + 12x = 0 \quad \rightarrow \quad x = 3$$

Three is not a root, because division by x–3 is division by zero.

4.1 Linear Equations

The general equation of the first degree in x and y is
(7) $ax+by+c=0$

We can solve for x or y. The solution for y is as follows.
(8a) $ax+by+c-(ax+c)=0-(ax+c)$ subtract $(ax+c)$ from both sides
(8b) $by=-ax-c$
(8c) $y=-\frac{a}{b}x-\frac{c}{b}$ fusince $b\neq 0$, divide both sides by b

On the other hand if a or b=0, then the equation reduces to one unknown, and the solution is straightforward.

(9) $ax+c=0 \quad (b=0) \rightarrow x=-\frac{c}{a}$ *if* $a\neq 0$*, can divide both sides by a*

(10) $by+c=0 \quad (a=0) \rightarrow y=-\frac{c}{b}$ *if* $b\neq 0$*, can divide both sides by b*

Algebraic solution of two equations There are two basic ways to find solutions: by elimination and by determinants. Solution by elimination takes two forms: (1) by addition or subtraction, (2) by substitution. For example solve the equations $3x+7y+9=0$ *and* $4x+5y-1=0$ for x and y. (As a practical matter the following process can be applied to 3 equations.)

1 Solution by addition or subtraction.

(11) $12x+28y+36=0$ *multiply by 4*

$-12x+15y-03=0$ *multiply by 3*

$13y+39=0$ *solve to get* $y=-3$

$4x+5y-1=4x-15-1=4x-16=0$ *solve to get* $x=4$

check $3x+7y+9=3(4)+7(-3)+9=0$ *qed*

check $4x+5y-1=4(4)+5(-3)-1=0$ *qed*

2 Solution by substitution.

(12) $x=\frac{1}{3}(-7y-9)$ *first equation*

$4\frac{1}{3}(-7y-9)+5y-1=0$ *substitute for x in second equation*

$-\frac{28}{3}y-12+5y-1=-\frac{13}{3}y-13=0$ *solve to get* $y=-3$

$x=\frac{1}{3}(-7y-9)=\frac{1}{3}(-7)(-3)-3=7-3=4$ *solve to get* $x=4$ *qed*

3 Solution by Determinants We manipulate two general simultaneous equations to show the origin of determinants.

(13*a*) $a_1x+b_1y+c_1=0$

(13*b*) $a_2x+b_2y+c_2=0$

$$(14)\quad \begin{array}{rl} b_2a_1x+b_2b_1y+b_2c_1=0 & \textit{multiply by } b_2 \\ \underline{-b_1a_2x-b_1b_2y-b_1c_2=0} & \textit{multiply by } -b_1 \\ (b_2a_1-b_1a_2)x+(b_2c_1-b_1c_2)=0 & \textit{add, solve for } x \end{array}$$

$$x=-\frac{(b_2c_1-b_1c_2)}{(a_1b_2-a_2b_1)} \quad \textit{multiply by } a_1,\ a_2 \textit{ to get } \quad y=-\frac{(a_2c_1-a_1c_2)}{(a_1b_2-a_2b_1)}$$

The symbol for a determinant is an $n\times m$ or $n\times n$ array of numbers bounded by bars (equation 15). The values for x and y in (14) are examples of *determinants* that are written in (16) as determinants in the form of columns of coefficients.

$$(15)\quad \begin{vmatrix} a_1 & b_1 \\ a_2 & b_2 \end{vmatrix}=a_1b_2-a_2b_1$$

From (13) $-c_1=a_1x+b_1y$ and $-c_2=a_2x+b_2y$

Then $\Delta=a_1b_2-a_2b_1$ and

$$(16)\quad x=\frac{\begin{vmatrix} -c_1 & b_1 \\ -c_2 & b_2 \end{vmatrix}}{\begin{vmatrix} a_1 & b_1 \\ a_2 & b_2 \end{vmatrix}}=\frac{-c_1b_2+c_2b_1}{\Delta} \qquad y=\frac{\begin{vmatrix} a_1 & -c_1 \\ a_2 & -c_2 \end{vmatrix}}{\begin{vmatrix} a_1 & b_1 \\ a_2 & b_2 \end{vmatrix}}=\frac{-a_1c_2+a_2c_1}{\Delta}$$

Substitute in (13) the numbers in $3x+7y+9=0$ *and* $4x+5y-1=0$ and solve for x and y.

(17*a*) $-c_1=a_1x+b_1y \quad\rightarrow\quad -9=3x+7y$

(17*b*) $-c_2=a_2x+b_2y \quad\rightarrow\quad 1=4x+5y$

$$\begin{vmatrix} c_1 & b_1 \\ c_2 & b_2 \end{vmatrix}=\begin{vmatrix} -9 & 7 \\ +1 & 5 \end{vmatrix}=-9\times5-7\times1=-52 \qquad \begin{vmatrix} a_1 & c_1 \\ a_2 & c_2 \end{vmatrix}=\begin{vmatrix} 3 & -9 \\ 4 & +1 \end{vmatrix}=3\times1-(-9)\times4=39$$

Then $\Delta=a_1b_2-a_2b_1=3\times5-7\times4=-13$ and

$$(18)\quad x=\frac{\begin{vmatrix} c_1 & b_1 \\ c_2 & b_2 \end{vmatrix}}{\begin{vmatrix} a_1 & b_1 \\ a_2 & b_2 \end{vmatrix}}=-\frac{c_1b_2-c_2b_1}{\Delta}=\frac{-52}{-13}=4 \quad y=\frac{\begin{vmatrix} a_1 & c_1 \\ a_2 & c_2 \end{vmatrix}}{\begin{vmatrix} a_1 & b_1 \\ a_2 & b_2 \end{vmatrix}}=\frac{a_1c_2-a_2c_1}{\Delta}=\frac{39}{-13}=-3$$

4.2 Quadratic Equations

The general quadratic equation is a *polynomial* of *degree* 2 where a, b, c are numbers of any kind, and whose roots are r_1, r_2.

(19) $ax^2+bx+c=0 \quad (a \neq 0)$

(20) $ax^2+bx+c=a(x-r_1)(x-r_2)$

Now

(21) $(x-r_1)(x-r_2)=x(x-r_2)-r_1(x-r_2)=x^2-(r_2+r_1)x+r_1r_2$

and

(22a) *if* $ax^2+bx+c=ax^2-a(r_2+r_1)x+ar_1r_2$

then equate coefficients

(22b) $b=-a(r_2+r_1), \qquad c=ar_1r_2$

(22c) $r_2+r_1=-\dfrac{b}{a}, \qquad r_1r_2=\dfrac{c}{a}$

> *If the roots of the quadratic equation*
> $ax^2+bx+c=0\,(a \neq 0)$ *are* r_1, r_2 *then* $r_2+r_1=-\dfrac{b}{a}$, $r_1r_2=\dfrac{c}{a}$
> *Conversely*
> *If* r_1, r_2 *are the roots of* $x^2+jx+k=0$ *then* $j=-(r_2+r_1)$, $k=r_1r_2$

Solution by Completing the Square The roots of any quadratic equation can be found by a *completing the square* process, which creates a polynomial that is the square of another polynomial. One learns in arithmetic that *multiplication is connected to addition* by the distributive law $x(y+z)=xy+xz$, which is used to derive the following.

(23) $(x+a)(x+a)=x(x+a)+a(x+a)=x^2+xa+ax+a^2$

so that $(x+a)^2=x^2+2ax+a^2$

let $m=2a$ *(see eqn23)* $\Rightarrow a=\dfrac{m}{2}$

Therefore to complete the square of x^2+mx *add* $a^2=\left(\frac{m}{2}\right)^2=\frac{m^2}{4}$

so that $x^2+mx+\frac{m^2}{4}=(x+\frac{m}{2})^2$

Example

(24*a*) *find the roots of* $2x^2-6x-5=0$

(24*b*) *add* 5 *to both sides* $2x^2-6x=5$

(24*c*) *divide by* 2 $x^2-3x=\frac{5}{2}$

(24*d*) *add* $\left(\frac{-3}{2}\right)^2=\frac{9}{4}$ *to both sides* $\rightarrow x^2-3x+\frac{9}{4}=\frac{5}{2}+\frac{9}{4} \rightarrow \left(x-\frac{3}{2}\right)^2=\frac{19}{4}$

(24*e*) $x-\frac{3}{2}=\pm\frac{\sqrt{19}}{2} \rightarrow x=\frac{3}{2}\pm\frac{\sqrt{19}}{2}$

Example

(25*a*) *find the roots of* $4w^2-7=-8w$

(25*b*) *add* $7+8w$ *to both sides* $4w^2+8w=7$

(25*c*) *divide by* 4 $w^2+2w=\frac{7}{4}$

(25*d*) *add* 1 *to both sides* $\rightarrow w^2+2w+1=\frac{7}{4}+1 \rightarrow (w+1)^2=\frac{11}{4}$

(25*e*) $w+1=\pm\frac{\sqrt{11}}{2} \rightarrow w=-1\pm\frac{\sqrt{11}}{2}$

Solve the pairs of linear equations algebraically, and check the results.

Problem 401 $2x-y+7=0 \quad 3x+4y-6=0$ (x,y=–2,3)

Problem 402 $2x-3y=10 \quad 2(x-10)-3(1+2y)=5-3x$ (x,y=8,2)

Problem 403 $3(2x-1)-5(2y+1)=0 \quad 5(3y-2)-2(5x-5)+15=0$

(x,y=3,1)

Solve by completing the square

Problem 404 $3x^2+11x=4$ x=–11/6±√(169/36)

Problem 405 $x^2+mx+n=0$ $x=0.5(-m\pm\sqrt{(m^2-4n)})$

Problem 406 $px(r-q)=(px)^2-qr$ x=(–q/p, r/p)

Solution by Quadratic Formula We express the roots as functions of coefficients a, b, c by applying the process of completing the square. To start transpose c and divide by a.

$$(26)\quad ax^2+bx+c=0 \;\rightarrow\; x^2+\frac{b}{a}x=-\frac{c}{a}$$

$$\textit{next add}\left(\frac{b}{2a}\right)^2=\frac{b^2}{4a^2}\ \textit{to both sides}$$

$$x^2+\frac{b}{a}x+\frac{b^2}{4a^2}=\frac{b^2}{4a^2}-\frac{c}{a}=\frac{b^2}{4a^2}-\frac{4ac}{4aa}$$

$$\left(x+\frac{b}{2a}\right)^2=\frac{b^2-4ac}{4a^2} \quad\rightarrow\quad x+\frac{b}{2a}=\pm\frac{\sqrt{b^2-4ac}}{2a}$$

$$x_1,x_2=-\frac{b}{2a}\pm\frac{\sqrt{b^2-4ac}}{2a}$$

Specific cases

$$(27)\quad \textit{If}\ c=0,\ \textit{then}\ ax^2+bx+c=ax^2+bx=ax(x+\frac{b}{a})\ \textit{and}\ x_1,x_2=0,-\frac{b}{a}$$

$$\textit{Check}\quad x_1,x_2=-\frac{b}{2a}\pm\frac{\sqrt{b^2-4ac}}{2a}=-\frac{b}{2a}\pm\frac{\sqrt{b^2-0}}{2a}=-\frac{b}{2a}\pm\frac{b}{2a}=0,-\frac{b}{a}\ \textit{qed}$$

$$(28)\quad \textit{If}\ b=0,\ \textit{then}\ ax^2+bx+c=ax^2+c=a(x^2+\frac{c}{a})\ \textit{and}\ x_1,x_2=\pm\sqrt{\frac{c}{a}}$$

$$\textit{Check}\quad x_1,x_2=-\frac{b}{2a}\pm\frac{\sqrt{b^2-4ac}}{2a}=-0\pm\frac{\sqrt{0-4ac}}{2a}=\pm\sqrt{\frac{4ac}{4a^2}}=\pm\sqrt{\frac{c}{a}}\quad \textit{qed}$$

Sum and product of the roots provide an easy way to check results.

$$(29)\quad x_1+x_2=-\frac{b}{2a}+\frac{\sqrt{b^2-4ac}}{2a}-\frac{b}{2a}-\frac{\sqrt{b^2-4ac}}{2a}=-\frac{b}{a}$$

$$(30)\quad x_1x_2=\frac{-b+\sqrt{b^2-4ac}}{2a}\times\frac{-b-\sqrt{b^2-4ac}}{2a}$$

$$=\frac{b^2-\left(\sqrt{b^2-4ac}\right)^2}{4a^2}=\frac{b^2-b^2+4ac}{4a^2}=\frac{c}{a}$$

Character of the roots

	roots of ax^2+bx+c *are*	*when*
(31*a*)	*real and unequal*	$b^2-4ac>0$
(31*b*)	*real and equal*	$b^2-4ac=0$
(31*c*)	*complex and unequal*	$b^2-4ac<0$

3 Examples - (take care, they are different)

(32a) $x^{-6}-7x^{-3}-8=0 \rightarrow \text{let } y=x^{-3} \rightarrow y^2-7y-8=0$

$$y_1, y_2 = -\frac{b}{2a} \pm \frac{\sqrt{b^2-4ac}}{2a} = -\frac{-7}{2} \pm \frac{\sqrt{49-4(-8)}}{2} = \frac{7}{2} \pm \frac{\sqrt{81}}{2} = \frac{7}{2} \pm \frac{9}{2} = -1, 8$$

(32b) $8=x^{-3} \rightarrow 8x^3-1=0 \rightarrow x_1=\frac{1}{2}$ is a root by inspection

factor $8x^3-1=(2x-1)(4x^2+2x+1)$

zeros of $4x^2+2x+1 \rightarrow x_2, x_3 = -\frac{2}{8} \pm \frac{\sqrt{4-16}}{8} = -\frac{1}{4} \pm i\frac{\sqrt{3}}{4}$

(32c) $-1=x^{-3} \rightarrow x^3+1=0 \rightarrow x_1=-1$ is a root by inspection

factor $x^3+1=(x+1)(x^2-x+1)$

zeros of $x^2-x+1 \rightarrow x_2, x_3 = -\frac{-1}{2} \pm \frac{\sqrt{1-4}}{2} = \frac{1}{2} \pm i\frac{\sqrt{3}}{2}$

Solve by quadratic formula:

Problem 407 $2x^2-11x=-12$ (3/2, 4)

Problem 408 $x-2x^2+15=0$ (−5/2, 3)

Problem 409 $2x^2+2x=1$ (1/2)(−1±√3)

Problem 410 $3x(x-1)+1=0$ (1/6)(3±i√3)

Problem 411 $24w-w^2=143$ (11,13)

Problem 412 $2mx=4mn-2nx+x^2$ (2m, 2n)

Problem 413 $a^2x-ax^2=ab-bx$ (a, b/a)

4.3 Determinant Operations

We use Cramer's rule (sidebar page 35) to solve the electric circuit node equations for v_2 and v_3. The two node solution is a good test of algebraic skills.

$$(33)\quad \begin{aligned} &node\,2 \quad i_S = y_{22}v_2 - y_{23}v_3 \\ &node\,3 \quad 0 = -y_{32}v_2 + y_{33}v_3 \qquad (general\ \ form) \end{aligned}$$

$$(34)\quad \Delta_Y = y_{22}y_{33} - y_{23}y_{32}$$

Figure 508 Low Pass Filter

$$(35a)\quad v_2 = \frac{\begin{vmatrix} i_S & -y_{23} \\ 0 & y_{33} \end{vmatrix}}{\Delta_Y} = \frac{i_S y_{33} - (-y_{23}\times 0)}{\Delta_Y}$$

$$(35b)\quad v_3 = \frac{\begin{vmatrix} y_{22} & i_S \\ -y_{32} & 0 \end{vmatrix}}{\Delta_Y} = \frac{(y_{22}\times 0) - (-y_{32}\times i_S)}{\Delta_Y}$$

Specific form let $R = R_1 = R_2,\ L = L_1 = L_2,\ C = C_1$

$$(36a)\quad node\,2 \quad \frac{1}{R+pL}v_S = \left(\frac{1}{R+pL} + pC + \frac{1}{pL}\right)v_2 \qquad -\frac{1}{pL}v_3$$

$$(36b)\quad node\,3 \qquad 0 = \qquad -\frac{1}{pL}v_2 + \left(\frac{1}{pL} + \frac{1}{R}\right)v_3$$

$$(37)\quad \Delta_Y = \left(\frac{1}{R+pL} + pC + \frac{1}{pL}\right)\left(\frac{1}{pL} + \frac{1}{R}\right) - \left(-\frac{1}{pL}\right)^2 = \frac{1}{pLR}\left(2 + pCR + p^2LC\right)$$

$$(38)\quad v_2 = \frac{\begin{vmatrix} i_S & -y_{23} \\ 0 & y_{33} \end{vmatrix}}{\Delta_Y} = \frac{\frac{1}{R+pL}v_S \cdot \left(\frac{1}{pL} + \frac{1}{R}\right)}{\frac{1}{pLR}\left(2 + pCR + p^2LC\right)} = \frac{v_S}{\left(2 + pCR + p^2LC\right)}$$

$$(39)\quad v_3 = \frac{\begin{vmatrix} y_{22} & i_S \\ -y_{32} & 0 \end{vmatrix}}{\Delta_Y} = \frac{\frac{1}{R+pL}v_S \cdot \frac{1}{pL}}{\frac{1}{pLR}\left(2 + pCR + p^2LC\right)} = \frac{Rv_S}{(R+pL)\left(2 + pCR + p^2LC\right)}$$

Problem 414 Derive equations 37, 38, and 39.

Algebraic solution by determinants We show the solution to equations 40 without proof.

$$(40a)\quad a_1x+b_1y+c_1z=d_1$$
$$(40b)\quad a_2x+b_2y+c_2z=d_2$$
$$(40c)\quad a_3x+b_3y+c_3z=d_3$$

Observe how the a, b, c columns are replaced by the d column in the solutions for x, y, z.

$$(41)\quad x=\frac{\begin{vmatrix} d_1 & b_1 & c_1 \\ d_2 & b_2 & c_2 \\ d_3 & b_3 & c_3 \end{vmatrix}}{\Delta}\qquad y=\frac{\begin{vmatrix} a_1 & d_1 & c_1 \\ a_2 & d_2 & c_2 \\ a_3 & d_3 & c_3 \end{vmatrix}}{\Delta}\qquad z=\frac{\begin{vmatrix} a_1 & b_1 & d_1 \\ a_2 & b_2 & d_2 \\ a_3 & b_3 & d_3 \end{vmatrix}}{\Delta}$$

where

$$(42)\quad \Delta=\begin{vmatrix} a_1 & b_1 & c_1 \\ a_2 & b_2 & c_2 \\ a_3 & b_3 & c_3 \end{vmatrix}$$

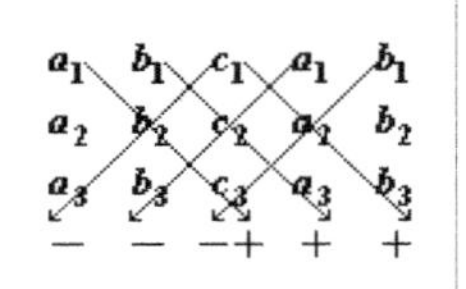

A fairly easy way to calculate a third order determinant, and only the third order, is to add columns 1 and 2 to the right of the determinant. The six products and their signs are indicated by the arrows.

$$(43)\quad \Delta=a_1b_2c_3+b_1c_2a_3+c_1a_2b_3-c_1b_2a_3-a_1c_2b_3-b_1a_2c_3$$

Given the equations x+2y–z=6, 2x–y+3z=–13, 3x–2y+3z=–16.

Problem 415 Solve the equations by substitution. Hint take two at a time. (x=–1, y=2, z=–3)

Problem 416 Solve the equations by addition or subtraction.

Given the equations 3p–2q+r=6, 2p+3q+2r=–1, 5q–4r=–3.

Problem 417 Solve the equations by substitution. Hint take two at a time. (p=3/2, q=–1, r=–1/2)

Problem 418 Solve the equations by addition or subtraction.

Cramer's Rule

The first subscript is the row number. The second subscript is the column number. Determinants are expanded by rows or columns.

Cramer's solutions are expansions by columns where forcing functions replace the column's elements. Note: incorporate minus signs into the a_{ij}'s.

Cramer found responses y_1, y_2 to forcing functions x_1, x_2

If $x_1 = a_{11}y_1 + a_{12}y_2$ and $x_2 = a_{21}y_1 + a_{22}y_2$

Then $\Delta = a_{11}a_{22} - a_{21}a_{12}$ and

$$y_1 = \frac{\begin{vmatrix} x_1 & a_{12} \\ x_2 & a_{22} \end{vmatrix}}{\begin{vmatrix} a_{11} & a_{12} \\ a_{21} & a_{22} \end{vmatrix}} = \frac{x_1a_{22} - x_2a_{12}}{\Delta} \qquad y_2 = \frac{\begin{vmatrix} a_{11} & x_1 \\ a_{21} & x_2 \end{vmatrix}}{\begin{vmatrix} a_{11} & a_{12} \\ a_{21} & a_{22} \end{vmatrix}} = \frac{x_2a_{11} - x_1a_{21}}{\Delta}$$

And for three responses y_1, y_2, y_3 *to forcing functions* x_1, x_2, x_3

If $x_1 = a_{11}y_1 + a_{12}y_2 + a_{13}y_3$

$x_2 = a_{21}y_1 + a_{22}y_2 + a_{23}y_3$

$x_3 = a_{31}y_1 + a_{32}y_2 + a_{33}y_3$

Then $\Delta = a_{11}\Delta_{11} - a_{21}\Delta_{21} + a_{31}\Delta_{31}$ *(expansion by column* 1)

$$\Delta = a_{11}(a_{22}a_{33} - a_{23}a_{32}) - a_{21}(a_{12}a_{33} - a_{13}a_{32}) + a_{31}(a_{12}a_{23} - a_{13}a_{22})$$

$$y_1 = \frac{x_1\Delta_{11} - x_2\Delta_{21} + x_3\Delta_{31}}{\Delta} \quad \textit{(expansion down column 1, rows 1, 2, 3)}$$

$$y_2 = \frac{x_1\Delta_{12} - x_2\Delta_{22} + x_3\Delta_{32}}{\Delta} \quad \textit{(expansion down column 2, rows 1, 2, 3)}$$

$$y_3 = \frac{x_1\Delta_{13} - x_2\Delta_{23} + x_3\Delta_{33}}{\Delta} \quad \textit{(expansion down column 3, rows 1, 2, 3)}$$

4.4 Solving Real Equations

We challenge the reader to start from equations 45 and finish with equations 61. Do the algebraic manipulations in equations 49 to 61 without looking at pages 36, 37, and 38.

Figure 401 Low Pass Filter

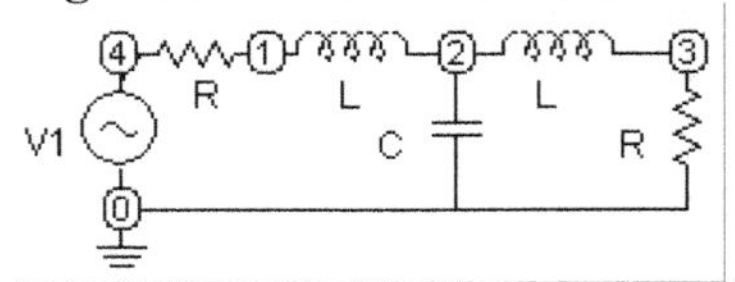

Polynomials originate in solutions to physical problems. For example Electrical Engineers analyze the low pass filter circuit and write equations 45a, 45b (Figure 401). Then the math takes over.

$$(45a)\quad node\,2\quad \frac{1}{R+pL}v_S = \left(\frac{1}{R+pL}+pC+\frac{1}{pL}\right)v_2 \qquad -\frac{1}{pL}v_3$$

$$(45b)\quad node\,3\quad 0 = \qquad -\frac{1}{pL}v_2+\left(\frac{1}{pL}+\frac{1}{R}\right)v_3$$

The R, L, C are real numbers in this, or any, electric circuit, because they represent physical devices. Real number coefficients result from analysis of electric circuits, analysis of bodies in motion, and analysis of all other physical problems. In algebra constants a, b, and c replace R, L, C.

Mathematicians start with general formats, which make a lot of sense. Here are equations 45a, 45b in general format. Emphasis - in physical problems the a_{jk}'s are real numbers.

$$(46a)\quad x_1 = a_{11}\,y_1 + a_{12}y_2$$

$$(46b)\quad x_2 = a_{21}y_1 + a_{22}\,y_2$$

In general simultaneous linear equations are solved by using Cramer's rule. The not very obvious solution is

$$(47)\quad \Delta = a_{11}a_{22} - a_{21}a_{12}$$

$$(48a)\quad y_1 = \frac{x_1a_{22}-x_2a_{12}}{\Delta} \qquad (48b)\quad y_2 = \frac{-x_1a_{21}+x_2a_{11}}{\Delta}$$

We compare equations 45 and 46 to relate the x' and a's to the RLC expressions.

$$(49)\quad \Delta = a_{11}a_{22}-a_{21}a_{12} = \left(\frac{1}{R+pL}+pC+\frac{1}{pL}\right)\left(\frac{1}{pL}+\frac{1}{R}\right)-\left(\frac{1}{pL}\right)\left(\frac{1}{pL}\right)$$

What follows is a real exercise in algebraic manipulations.

The Arithmetic distribution law carries over to algebra. The law connects multiplication to addition.

(50) $x(y+z)=xy+xz$

Apply the distribution law once to get

(51) $\Delta=\left(\frac{1}{R+pL}+pC+\frac{1}{pL}\right)\left(\frac{1}{pL}\right)+\left(\frac{1}{R+pL}+pC+\frac{1}{pL}\right)\left(\frac{1}{R}\right)-\left(\frac{1}{pL}\right)\left(\frac{1}{pL}\right)$

Apply it again to get

(52) $\Delta=\left(\frac{1}{(R+pL)(pL)}+\frac{pC}{pL}+\frac{1}{p^2L^2}\right)+\left(\frac{1}{(R+pL)R}+\frac{pC}{R}+\frac{1}{pLR}\right)-\left(\frac{1}{p^2L^2}\right)$

Observe that the $1/p^2L^2$ terms cancel. Drop the parentheses.

(53) $\Delta=\frac{1}{(R+pL)(pL)}+\frac{pC}{pL}+\frac{1}{(R+pL)R}+\frac{pC}{R}+\frac{1}{pLR}$

Multiply *both* sides of (53) by (R+pL)(pLR), apply the distribution law, and cancel terms (the operations may not be clear at this point).

(54) $(R+pL)(pLR)\Delta=R+(R+pL)(R)(pC)+pL+(R+pL)(pL)(pC)+(R+pL)$

Add the R and pL terms

(55) $(R+pL)(pLR)\Delta=(R+pL)+(R+pL)(R)(pC)+(R+pL)(pL)(pC)+(R+pL)$

Factor out the (R+pL) term from each of the four terms. Think of this as a reverse distribution.

(56) $(R+pL)(pLR)\Delta=(R+pL)[1+(R)(pC)+(pL)(pC)+1]$

Divide both sides by (R+pL)(pLR)

(57) $\Delta=\frac{1}{pLR}[1+(R)(pC)+(pL)(pC)+1]=\frac{1}{pLR}[2+pCR+p^2LC]$

Here is a copy of the equations

$$(45a)\quad node\,2\quad \frac{1}{R+pL}v_S = \left(\frac{1}{R+pL}+pC+\frac{1}{pL}\right)v_2 - \frac{1}{pL}v_3$$

$$(45b)\quad node\,3\quad 0 = -\frac{1}{pL}v_2 + \left(\frac{1}{pL}+\frac{1}{R}\right)v_3$$

$$(47)\quad \Delta = a_{11}a_{22} - a_{21}a_{12}$$

$$(48a)\quad y_1 = \frac{x_1a_{22}-x_2a_{12}}{\Delta} \qquad (48b)\quad y_2 = \frac{-x_1a_{21}+x_2a_{11}}{\Delta}$$

Form the y_1 numerator.

$$(58)\quad x_1a_{22}-x_2a_{12} = \frac{1}{R+pL}v_S\left(\frac{1}{pL}+\frac{1}{R}\right) - 0\cdot\left(-\frac{1}{pL}\right)$$

Add the 2 fractions the same way one would in Arithmetic.

$$(59)\quad \left(\frac{1}{pL}+\frac{1}{R}\right) = \frac{1}{pL}\frac{R}{R}+\frac{1}{R}\frac{pL}{pL} = \frac{R+pL}{pLR}$$

Substitute 59 in equation 58, and form the y_1, y_2 numerators.

$$(60a)\quad x_1a_{22}-x_2a_{12} = \frac{1}{R+pL}v_S\frac{R+pL}{pLR} - 0\cdot\left(-\frac{1}{pL}\right) = \frac{v_S}{pLR}$$

$$(60b)\quad -x_1a_{21}+x_2a_{11} = -\frac{1}{R+pL}v_S\left(-\frac{1}{pL}\right) - 0 = \frac{v_S}{(R+pL)pL}$$

Apply Cramer's Rule

$$(61a)\quad v_2 = \frac{x_1a_{22}-x_2a_{12}}{\Delta_Y} = \frac{\frac{1}{pLR}v_S}{\frac{1}{pLR}\left(2+pCR+p^2LC\right)} = \frac{v_S}{\left(2+pCR+p^2LC\right)}$$

$$(61b)\quad v_3 = \frac{-x_1a_{21}+x_2a_{11}}{\Delta_Y} = \frac{\frac{1}{(R+pL)pL}v_S}{\frac{1}{pLR}\left(2+pCR+p^2LC\right)}$$

$$\frac{v_3}{v_S} = \frac{R}{(R+pL)\left(2+pCR+p^2LC\right)}$$

4.5 Elimination Operations

This non-trivial example shows the significant role of elimination by addition or subtraction, and by substitution in solving equations. EE dependant sources circuit analysis produces equations 62a to 62f. The v's and i's are variables, the R's and g_m are constants. Algebraic analysis solves for the signal transmission T = v_3/v_1 from input v_1 to output v_3.

Figure 408 V to I Circuit

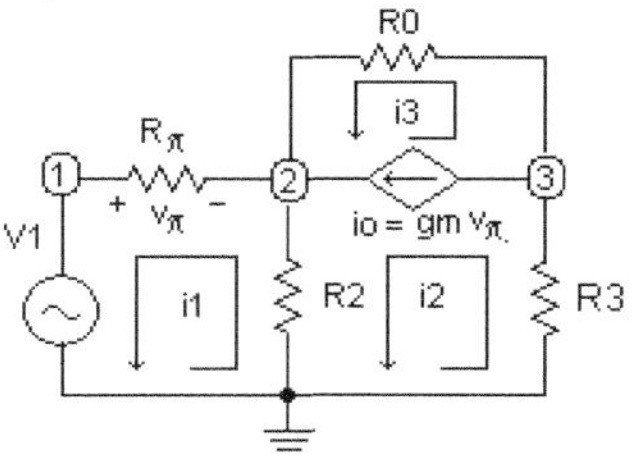

Again we challenge the reader to start from equation 62a and finish with equation 62g.

$$(62a)\quad v_1 = i_1 R_\pi + (i_1 - i_2) R_2 \qquad mesh\ 1$$
$$(62b)\quad 0 = (i_2 - i_1) R_2 + v_x + i_2 R_3 \quad mesh\ 2$$
$$(62c)\quad 0 = -v_x + i_3 R_0 \qquad mesh\ 3$$
$$(62d)\quad i_0 = g_m v_\pi = i_2 - i_3$$
$$(62e)\quad v_\pi = i_1 R_\pi$$
$$(62f)\quad v_3 = i_2 R_3 \quad \rightarrow \quad T = \frac{v_3}{v_1} = \frac{i_2 R_3}{v_1} = R_3 \frac{i_2}{v_1}$$

We need an expression for i_2/v_1 (62f). Voltage v_1 in equation 62a is a function of i_1 and i_2. We need to eliminate i_1 so that v_1 is only a function of i_2. Then we have a solution for transmission T (62f).

We do not need v_x so add 62c to 62b to eliminate the v_x *terms*.

$$(62b)\qquad (i_2 - i_1) R_2 + v_x + i_2 R_3 \qquad mesh\ 2$$
$$(62c)\qquad -v_x \qquad + i_3 R_0 \quad mesh\ 3$$
$$(62b + c)\quad (i_2 - i_1) R_2 + 0 \ + i_2 R_3 + i_3 R_0 \quad mesh\ 2+3$$

Replace 62b and 62c with 62b+c

$$(62a)\qquad v_1 = i_1 R_\pi + (i_1 - i_2) R_2 \qquad mesh\ 1$$
$$(62b + c)\quad 0 = (i_2 - i_1) R_2 + i_2 R_3 + i_3 R_0 \quad mesh\ 2+3$$
$$(62d)\quad g_m v_\pi = i_0 = i_2 - i_3$$

Use the distribution law to collect terms in 62a and 62b+c so that currents are *factors* as in $i_1 R_\pi + i_1 R_2 = i_1 (R_\pi + R_2)$

$$(64a)\quad v_1 = i_1 (R_\pi + R_2) - i_2 R_2 \qquad mesh\ 1$$
$$(64b)\quad 0 = -i_1 R_2 + i_2 (R_2 + R_3) + i_3 R_0 \quad mesh\ 2+3$$
$$(62d)\quad g_m v_\pi = i_0 = i_2 - i_3$$

(62*d*) $g_m v_\pi = i_2 - i_3$

(65*a*) *add* i_3 *to both sides* $i_3 + g_m v_\pi = i_2 - i_3 + i_3 \rightarrow i_3 + g_m v_\pi = i_2$

(65*b*) *sub* $g_m v_\pi$ *from both sides* $i_3 + g_m v_\pi - g_m v_\pi = i_2 - g_m v_\pi \rightarrow i_3 = i_2 - g_m v_\pi$

(65*c*) *sub* $i_1 R_\pi$ *for* $v_\pi \rightarrow i_3 = i_2 - g_m i_1 R_\pi$

Now substitute the i_3 equation 65c for i_3 in 64b.

(66*a*) $v_1 = i_1(R_\pi + R_2) - i_2 R_2$ *mesh 1*

(66*b*) $0 = -i_1 R_2 + i_2(R_2 + R_3) + (i_2 - g_m R_\pi i_1) R_0$ *mesh 2+3*

Collect terms again.

(67*a*) $v_1 = i_1(R_\pi + R_2) - i_2 R_2$ *mesh 1*

(67*b*) $0 = -i_1(R_2 + g_m R_\pi R_0) + i_2(R_0 + R_2 + R_3)$ *mesh 2+3*

Equation 67a shows v_1 depends on i_1 and i_2, and equation 67b relates i_1 to i_2. In equation 62f we need i_2/v_1 so eliminate i_1 by substitution.

(67b) $0 = -\mathrm{i}_1(\mathrm{R}_2 + \mathrm{g}_\mathrm{m}\mathrm{R}_\pi\mathrm{R}_0) + \mathrm{i}_2(\mathrm{R}_0 + \mathrm{R}_2 + \mathrm{R}_3)$ mesh 2+3

(67c) move i_1 term to left side of 64b $\mathrm{i}_1(\mathrm{R}_2 + \mathrm{g}_\mathrm{m}\mathrm{R}_\pi\mathrm{R}_0) = \mathrm{i}_2(\mathrm{R}_0 + \mathrm{R}_2 + \mathrm{R}_3)$

(67d) divide both sides by $(\mathrm{R}_2 + \mathrm{g}_\mathrm{m}\mathrm{R}_\pi\mathrm{R}_0) \rightarrow \mathrm{i}_1 = \mathrm{i}_2 \dfrac{(\mathrm{R}_0 + \mathrm{R}_2 + \mathrm{R}_3)}{(\mathrm{R}_2 + \mathrm{g}_\mathrm{m}\mathrm{R}_\pi\mathrm{R}_0)}$

Substitute 67d into 67a to eliminate i_1.

$$(68) \quad v_1 = i_2 \frac{(R_0 + R_2 + R_3)}{(R_2 + g_m R_\pi R_0)}(R_\pi + R_2) - i_2 R_2$$

Substitute the reciprocal of v_1/i_2 into 62f.

$$(62f) \quad \frac{v_3}{v_1} = \frac{i_2}{v_1} R_3 = \frac{1}{\dfrac{(R_0 + R_2 + R_3)}{(R_2 + g_m R_\pi R_0)}(R_\pi + R_2) - R_2} R_3$$

The form of this result can be improved by multiplying numerator and denominator by $(R_2 + g_m R_\pi R_0)$ and multiplying out the denominator.

$$(62g) \quad \frac{v_3}{v_1} = \frac{i_2}{v_1} R_3 = \frac{(R_2 + g_m R_\pi R_0) R_3}{(R_0 + R_2 + R_3)(R_\pi + R_2) - (R_2 + g_m R_\pi R_0) R_2}$$

5 Exponents

An *exponent n* is a symbol written above, and on the right of, another symbol known as the *base x* as in x^n. Another exponent form is b^x with base b and variable x as the exponent (Chapter 7).

The expression x^n is referred to as an *algebraic power function*; specifically the nth power of x. All arithmetic operations apply to exponents. *The base and exponent can be any type of numbers*. However here we restrict exponents to real numbers.

5.1 Positive Integer Exponents

The symbol x^n is read as "x to the nth power" or "x to the nth", or whatever you choose to say (x^2 is x squared, x^3 is x cubed). When n is a positive integer the world says *exponent n is the nth power of base x*. And, when n is an integer, *the symbol x^n represents the product of n factors each equal to x*.

(1) $x^n = x \cdot x \cdot x \cdots x \cdot x, \ n \ factors \ x$

Laws of Positive Integer Exponents The laws of Positive Integer exponents extend algebraic processes to include *powers* of variables.

Addition of powers

(2) $x^n \cdot x^m = x^{n+m}$

$$Proof \ x^n \cdot x^m = (x \cdot x \cdots x, \ n \ factors \ x)(x \cdot x \cdots x, \ m \ factors \ x)$$
$$= x \cdot x \cdots x, \ n+m \ factors \ x$$
$$= x^{n+m}$$

$Example \ 3^5 7^2 3^3 = 3^8 7^2 \ \ exponents \ add \ only \ when \ bases \ are \ equal$

Power of a power of x

(3) $(x^m)^n = x^{mn}$

$$Proof \ (x^m)^n = x^m \cdot x^m \cdots x^m, n \ factors \ x^m$$
$$= x^{m+m+\cdots+m \ n \ terms \ m} = x^{mn}$$

$Example \ (4^3)^8 = 4^{24} \ \ this \ is \ a \ product \ of \ eight \ 4^3 \ where \ eight \ 3's \ add$

Power of a product of powers of x and y

$$(4) \quad (xy)^n = x^n y^n$$

$$\textit{Proof} \ (xy)^n = xy \cdot xy \cdots xy, \ n \ \textit{factors} \ xy$$
$$= (x \cdot x \cdots x, \ n \ \textit{factors} \ x)(y \cdot y \cdots y, \ n \ \textit{factors} \ y)$$
$$= x^n y^n$$

$$\textit{Example} \ (2\times 9)^5 = 2^5 9^5$$

$\textit{product of five } 2\times 9\text{'s is product of five } 2\text{'s \& five } 9\text{'s}$

Power of a fraction of powers of x and y

$$(5) \quad \left(\frac{x}{y}\right)^n = \frac{x^n}{y^n}$$

$$\textit{Proof} \ \left(\frac{x}{y}\right)^n = \left(\frac{x}{y}\right)\cdot\left(\frac{x}{y}\right)\cdots\left(\frac{x}{y}\right), \ n \ \textit{factors} \ \left(\frac{x}{y}\right)$$
$$= \frac{x \cdot x \cdots x, \ n \ \textit{factors} \ x}{y \cdot y \cdots y, \ n \ \textit{factors} \ y}$$
$$= \frac{x^n}{y^n}$$

$$\textit{Example} \ \left(\frac{5}{6}\right)^8 = \left(\frac{5}{6}\right) \textit{multiplied 8 times} = \frac{5 \ \textit{multiplied 8 times}}{6 \ \textit{multiplied 8 times}} = \frac{5^8}{6^8}$$

Power of a ratio of powers of x where m>n

$$(6) \quad \frac{x^m}{x^n} = x^{m-n} \quad \textit{where} \ m > n, \quad x \neq 0$$

$$\textit{Proof} \ \frac{x^m}{x^n} = \frac{x \cdot x \cdots x, \ m \ \textit{factors} \ x}{x \cdot x \cdots x, \ n \ \textit{factors} \ x}$$
$$= \frac{(x \cdot x \cdots x, \ m-n \ \textit{factors} \ x)(x \cdot x \cdots x, \ n \ \textit{factors} \ x)}{(x \cdot x \cdots x, \ n \ \textit{factors} \ x)}$$
$$= x \cdot x \cdots x, \ m-n \ \textit{factors} \ x = x^{m-n}$$

$$\textit{Example} \ \frac{4^7}{4^5} = \frac{4^7}{4^5}\frac{4^{-5}}{4^{-5}} = \frac{4^7 \times 4^{-5}}{4^5 \times 4^{-5}} = \frac{4^2}{4^0} = 4^2$$

Power of a ratio of powers of x where n>m

$$(7)\quad \frac{x^m}{x^n} = \frac{1}{x^{n-m}} \quad \text{where } n > m,\ x \neq 0$$

$$\textit{Proof}\quad \frac{x^m}{x^n} = \frac{x \cdot x \cdots x,\ m \text{ factors } x}{x \cdot x \cdots x,\ n \text{ factors } x}$$

$$= \frac{(x \cdot x \cdots x,\ m \text{ factors } x)}{(x \cdot x \cdots x,\ n-m \text{ factors } x)(x \cdot x \cdots x,\ m \text{ factors } x)}$$

$$= \frac{1}{x \cdot x \cdots x,\ n-m \text{ factors } x} = \frac{1}{x^{n-m}}$$

$$\textit{Example}\quad \frac{4^3}{4^7} = \frac{4^3}{4^7}\frac{4^{-3}}{4^{-3}} = \frac{4^3 \times 4^{-3}}{4^7 \times 4^{-3}} = \frac{4^0}{4^4} = \frac{1}{4^4}$$

Examples of Positive Integer Exponents

$$(8)\quad x^4 \cdot x^{13} = x^{17} \qquad (x^5)^7 = x^{35} \qquad (xy)^{12} = x^{12}y^{12}$$

$$\left(\frac{x}{y}\right)^8 = \frac{x^8}{y^8} \qquad \frac{x^7}{x^5} = x^2 \qquad \frac{x^3}{x^{13}} = \frac{1}{x^{10}}$$

5.2 Fractional Exponents

Let m=1/q where q is any positive integer. If 1/q, as an exponent, is to obey the addition law, then

$$(9)\quad \left(x^{\frac{1}{q}}\right)^q = x^{\frac{1}{q}} \cdot x^{\frac{1}{q}} \cdots x^{\frac{1}{q}} = x^{\frac{1}{q}+\frac{1}{q}+\cdots+\frac{1}{q}\ \ \text{q terms}\frac{1}{q}} = x^{\frac{q}{q}} = x^1 = x$$

$\therefore\ x^{\frac{1}{q}}$ is defined as the qth root of x

Example $\left(x^{\frac{1}{6}}\right)^6$ = product of six $\left(x^{\frac{1}{6}}\right) = x$ because sum of six $\frac{1}{6}$ equals 1

Changing from q to p factors we get

$$(10)\quad \left(x^{\frac{1}{q}}\right)^p = x^{\frac{1}{q}} \cdot x^{\frac{1}{q}} \cdots x^{\frac{1}{q}}[\text{p factors } x^{\frac{1}{q}}] = x^{\frac{1}{q}+\frac{1}{q}+\cdots+\frac{1}{q}\ \ \text{p terms}\frac{1}{q}} = x^{\frac{p}{q}}$$

Example $\left(x^{\frac{1}{6}}\right)^5$ = product of five $\left(x^{\frac{1}{6}}\right) = x^{\frac{5}{6}}$

because sum of five $\frac{1}{6}$ equals $\frac{5}{6}$

Changing the exponent to p/q we get

(11) $\left(x^{\frac{p}{q}}\right)^q = x^{\frac{p}{q}} \cdot x^{\frac{p}{q}} \cdot \cdots x^{\frac{p}{q}}, \text{q factors } x^{\frac{p}{q}} = x^{\frac{p}{q}+\frac{p}{q}+\cdots+\frac{p}{q} \text{ q terms } \frac{p}{q}} = x^{q\frac{p}{q}} = x^p$

Example $\left(x^{\frac{2}{7}}\right)^7$ = product of seven $\left(x^{\frac{2}{7}}\right) = x^2$

because sum of seven $\frac{2}{7}$ equals 2

We use fractional exponents to avoid radicals such as $\sqrt[3]{2^5}$

We avoid radicals like the plague, because radicals are an easy source of errors whereas fractional exponents are safe.

5.3 Exponent Zero

If zero, as an exponent, is to obey the addition law, then when x≠0

(12) $x^n x^0 = x^{n+0} = x^n \quad \Rightarrow \quad 1 = \frac{x^n}{x^n} = x^{n-n} = x^0 \quad \Rightarrow \quad \underline{x^0 = 1}$

Example since $\frac{a}{a} = 1$

$$\frac{10^3}{10^3} = \frac{10^3}{10^3} \times \frac{10^{-3}}{10^{-3}} = \frac{10^{3-3}}{10^0} = 10^{3-3} = 10^0 = 1$$

5.4 Negative Exponents

(13) $x^n x^{-n} = x^{n-n} = x^0 = 1 \quad \Rightarrow \quad x^{-n} = \frac{1}{x^n}$

Example $\frac{1}{x^5} = \frac{1}{x^5} \cdot \frac{x^{-5}}{x^{-5}} = \frac{x^{-5}}{x^0} = x^{-5}$

Problems 5

Simplify

1. $2^8 4^5$ 2. $27^5/3^{11}$ 3. $25^{x+2}/5^{x-1}$ 4. $9^{2m}(3^m)^{m+1}$ 5. $4^2 2^{3n}/8^{n+2}$

6. $\dfrac{c^{x^2}}{c^{x^2(x+1)}}$ 7. $\dfrac{(a^{2x-y})^{x+2y}}{(a^{2x+y})^{x-2y}}$ 8. $\dfrac{x^{(a^2-9)}}{x^{a-3}}$ 9. $\dfrac{a^{m-2n}a^{3(m+n)}}{a^{2m-n}}$ 10. $\left(\dfrac{b^{2x-3}}{b^{2x+3}}\right)^{x+1}$

Find the values. Hint write negative numbers –n as –1×n.

1. $81^{\frac{1}{2}}$ 2. 81^0 3. $0^{\frac{1}{2}}$ 4. $64^{\frac{1}{4}}$ 5. $27^{\frac{1}{3}}$ 6. $27^{\frac{2}{3}}$
7. $27^{\frac{4}{3}}$ 8. $16^{\frac{1}{4}}$ 9. $16^{\frac{3}{4}}$ 10. $\left(\frac{9}{25}\right)^{\frac{1}{2}}$ 11. $\left(\frac{9}{25}\right)^{\frac{3}{2}}$ 12. $0.04^{\frac{1}{2}}$
13. $0.216^{\frac{2}{3}}$ 14. $(-8)^{\frac{1}{3}}$ 15. $\left(-\frac{1}{32}\right)^{\frac{2}{5}}$ 16. $(-4)^3$ 17. 7^{-2} 18. $\left(\frac{2}{3}\right)^{-3}$

Convert to positive exponents.

1. x^{-2} 2. $x^{\frac{3}{4}}x^{-\frac{1}{2}}$ 3. $(x^{-\frac{1}{2}})^{-\frac{5}{3}}$ 4. $(x^{-\frac{1}{2}})^{-\frac{5}{3}}$ 5. $(-x^{-\frac{5}{6}})^{-\frac{1}{5}}$

6. $2x^{-1}y^{-2}$ 7. $\dfrac{3x^{-3}}{yz^{-4}}$ 8. $\dfrac{2x^{-1}y^4}{3^{-2}x^3y^{-5}}$ 9. $\dfrac{2^{-1}b^3c^{-\frac{2}{3}}}{5b^{-\frac{1}{4}}c^2}$ 10. $\dfrac{3x^{-\frac{2}{5}}y^{-\frac{3}{2}}}{2^{-2}x^{-\frac{1}{2}}y^{-\frac{5}{6}}}$

Convert denominator to 1.

1. $\dfrac{3x^2}{z^{-3}}$ 2. $\dfrac{3a}{x^4z^{-3}}$ 3. $\dfrac{x^2}{4y^{-\frac{2}{3}}}$ 4. $\dfrac{x^{(a^2-9)}}{x^{a-3}}$ 5. $\dfrac{c^{x^2}}{c^{x^2(x+1)}}$

Simplify

1. $x^{-1}-y^{-1}$ 2. $(x^{-3}+y^{-3})^{-1}$ 3. $\dfrac{1}{x^{-2}-y^{-2}}$

4. $\dfrac{x^{-1}+y^{-1}}{y^{-2}-x^{-2}}$ 5. $\dfrac{xy^{-2}+x^{-2}y}{x^{-1}-y^{-1}}$ 6. $\left(\dfrac{xy^{-1}-x^{-1}y}{xy^{-2}-x^{-2}y}\right)^{-1}$

7. $(x^{\frac{2}{3}}-y^{\frac{2}{3}})^3$ 8. $\dfrac{(a^2+x^2)^{-\frac{3}{5}}-x^{-2}}{(a^2+x^2)^{\frac{3}{5}}-x^2}$ 9. $\dfrac{3+9x(9x^2+1)^{-\frac{1}{2}}}{3x+(9x^2+1)^{\frac{1}{2}}}$

10. $\dfrac{(1-4x^2)^{\frac{1}{2}}-2x}{4x(1-4x^2)^{-\frac{1}{2}}-2}$ 11. $\dfrac{\frac{1}{5}x^{-2}-x^{-1}}{(x\div\frac{1}{125}x^{-2})^{-\frac{2}{3}}-1}$ 12. $\dfrac{x^2(x^2-a^2)^{-\frac{1}{2}}-(x^2-a^2)^{-\frac{1}{2}}}{(x^2-a^2)}$

6 The Binomial Theorem for any Index

The very useful Binomial Theorem shows how to expand $(x+a)^n$ when n is an integer thereby avoiding the tedious process of multiplying by $(x+a)$ n–1 times.

Furthermore, the Binomial Theorem shows how to expand $(x+a)^n$ when n is any number, positive, negative, integral or fractional where x and a can be any numbers. The Binomial Theorem has many applications such as calculating $(1.08)^4$ when formed as $(1+0.08)^4$ to 5 significant figures.

6.1 Product of Factors (n integer)

We know how to use the distributive law to multiply factors such as

(1) $(x+a)(x+b)=x(x+b)+a(x+b)=x^2+xb+ax+ab$

When we multiply by a third factor we get

(2) $(x+a)(x+b)(x+c)=x^3+(a+b+c)x^2+(ab+bc+ca)x+abc$

Actual multiplication is avoided if we proceed another way.
(1) Take x from 3 factors. The product is x^3.
(2) Take x from 2 factors and the constant from the third factor in 3 ways to get ax^2, bx^2, cx^2.
(3) Take x from 1 factor and the constant from 2 factors in 3 ways to get abx, bcx, cax.
(4) Take the constant from 3 factors in 1 way to get abc.
The sum of these terms confirms (2).

A particular case of equation 2 is produced by letting $a=b=c$.

(3) $(x+a)^3=x^3+3x^2a+3xa^2+a^3$

In the same way we can show that

(4) $(x+a)^4=x^4+4x^3a+6x^2a^2+4xa^3+a^4$

Avoiding multiplying out is an easier way to expand $(x+a)^n$ when n is an integer by asking the question “how many ways to choose x from a group of factors?” This brings us to permutations and combinations.

6.2 Permutations

On many occasions we need to know the number of different ways events can occur. Events such as dealing a poker hand, selecting of 2 out of 7 items, and our immediate concern, how many ways to choose x and a from a group of factors. The two basic kinds of different ways to do these operations are permutations and combinations.

Three letters *abc* taken 3 at a time are written as *abc, acb, bac, bca, cab, cba,* each of which is referred to as an *order*. Thus the 3 letters *abc* can be written down in 6 different orders. Chose 1 of 3 letters for the first position, 1 of 2 for the second position, and 1 of 1 for the third position. I.e. *abc, acb, bac, bca, cab, cba. The sequence of letters defines the order*. Clearly there are 3×2×1 ways to fill the positions. The number 3×2×1 is referred to as 3! (3 factorial), 4×3×2×1 is referred to as 4!, and so forth.

Definitions *If* n *is a positive integer, then* $n!$ *(n fractorial) is defined as*

$$n! = n(n-1)(n-2)(n-3)\cdots 3\cdot 2\cdot 1 \quad and \quad 0! = 1$$

Example $5! = 5\cdot 4\cdot 3\cdot 2\cdot 1 = 120$

Theorem 1 *Given* k *positions, numbered 1, 2,...., k, and* n *letters* $(n \geq k)$, *there are* $n(n-1)(n-2)....(n-\{k-1\})$ *different ways of assigning* k *of the* n *letters to the* k *positions*

Example *If* $n=4,\ k=3,$ *then* $n-(k-1) = 4-2 = 2$ *and* $(4)(3)(2) = 24$

The number of different ways is the number of permutatios ${}_nP_k$ of n items taken k at a time

$$(5)\quad {}_nP_k = n(n-1)(n-2)....(n-\{k-1\})\times\frac{(n-k)!}{(n-k)!} = \frac{n!}{(n-k)!}$$

Example ${}_6P_2 = 6(6-\{2-1\}) = 6(6-1) = \dfrac{6!}{(6-2)!} = \dfrac{6\cdot 5\cdot 4\cdot 3\cdot 2\cdot 1}{4\cdot 3\cdot 2\cdot 1} = 6\cdot 5 = 30$

The example shows that the factorial form is not necessary. However the factorial form provides a compact formula for ${}_nP_k$.

6.3 Combinations

Three letters *abc* taken 2 at a time are written *without regard to order* as *ab, ac, bc.* The different ways are referred to as *combinations*.

Three letters *abc* taken 3 at a time can be written down *without regard to order* in 1 way *abc*.

Theorem 2 Given n letters, the number of different ways to select k of the n letters $(n \geq k)$ *with no regard to order is*

$$_nC_k = \frac{n(n-1)(n-2)....(n-\{k-1\})}{k!} = \frac{_nP_k}{k!}$$

Proof Let $_nC_k$ be the number of different ways to select without regard to order. Any one selection of k letters can be arranged in $k!$ different orders. Therefore there are $k! \times {_nC_k}$ ways to fill k positions. By Theorem 1 the number of different ways to select *with* regard to order is $_nP_k$. Then we have $k! \times {_nC_k} = {_nP_k}$ so that $_nC_k = {_nP_k} / k!$.

Definition

The number of ways in which k letters can be selected from n letters $(n \geq k)$*, with no regard to order, is referred to as the number of combinations* $_nC_k$ *of n letters taken k at a time*

The number of combinatios

$$(6) \quad _nC_k = \frac{n(n-1)(n-2)....(n-\{k-1\})}{k!} \times \frac{(n-k)!}{(n-k)!} = \frac{n!}{k!(n-k)!}$$

Example $\quad {_6C_2} = \frac{6!}{2!(6-2)!} = \frac{6 \cdot 5 \cdot 4 \cdot 3 \cdot 2 \cdot 1}{2 \cdot 1 \cdot 4 \cdot 3 \cdot 2 \cdot 1} = \frac{6 \cdot 5}{2} = 15$

There are a multitude of relations involving the $_nC_k$. One such relation applicable to the Binomial Theorem is

$$(7) \quad _nC_k = \frac{n!}{k!(n-k)!} = \frac{n!}{(n-k)!k!} = {_nC_{n-k}}$$

6.4 The Binomial Theorem (integer index)

Binomial Theorem (*n integer*) *When n is a positive integer*

$$(x+a)^n = x^n + {}_nC_1 x^{n-1}a^1 + ... + {}_nC_k x^{n-k}a^k + ... + {}_nC_{n-1}x^1 a^{n-1} + a^n$$

where ${}_nC_k = \dfrac{n!}{k!(n-k)!}$ *so that*

$$(x+a)^n \equiv x^n + nx^{n-1}a^1 + \frac{n(n-1)}{2!}x^{n-2}a^2 + ... + \frac{n!}{k!(n-k)!}x^{n-k}a^k + ... + nx^1a^{n-1} + a^n$$

Proof Consider the product $(x+a)(x+a)(x+a)\ldots(x+a)$ *n factors*

The product is the sum of all the products we can form by selecting one term from each parens and multiplying.

(1) Take x from each of n parens, and multiply. The term is x^n.

(2) Take a from one paren, and x from each of n–1 parens, and multiply. the term is $x^{n-1}a$.

(3) With $k<n$ we select k out of n parens in ${}_nC_k$ ways. From the k parens we take a out of each of them, and we take x out of the remaining $n-k$ parens. The product is $x^{n-k}a^k$. Therefore the term is ${}_nC_k x^{n-k}a^k$.

(4) Take a from each of n parens, and multiply. The term is a^n.

Thus the sum of all the terms we get by multiplying out the parens is

$$x^n + {}_nC_1 x^{n-1}a^1 + ... + {}_nC_k x^{n-k}a^k + ... + {}_nC_{n-1}x^1a^{n-1} + a^n$$

The ${}_nC_k$ are referred to as the binomial coefficients. The name binomial is a carry over from the technical term binomial for $x+y$.

Expand by the binomial theorem (calculating the ${}_nC_k$.)

Problem 601 $(a+b)^4$

Problem 602 $\left(\frac{1}{2}b - 3x^3\right)^3$

Problem 603 $(e^x + e^{-x})^9$

6.5 Factorial n for any Index

Factorial n terminates only when n is a positive integer. Factorial n does *not* terminate when n is a negative integer or a fraction.

Definition If n is a positive integer, then n! is defined as

$$n! = n(n-1)(n-2)(n-3)....3\cdot 2\cdot 1$$

and $0! = 1$, because $n! = n(n-1)!$ so that $1 = 1! = 1(1-1)! = 0!$

(8) *If n is a positive integer, then n! terminates. The last number is 1*

$$7! = 7(7-1)(7-2)(7-3)(7-4)(7-5)(7-6) = 7\times 6\times 5\times 4\times 3\times 2\times 1$$

(9) *If n is a negative integer, then n! does not terminate*

$$-3! = -3(-3-1)(-3-2)(-3-3).... = -3\times -4\times -5\times -6\times ....$$

(10) *If n is a fraction, then n! does not terminate*

$$\frac{1}{2}! = \frac{1}{2}\left(\frac{1}{2}-1\right)\left(\frac{1}{2}-2\right)\left(\frac{1}{2}-3\right)\cdots = \frac{1}{2}\times -\frac{1}{2}\times -\frac{3}{2}\times -\frac{5}{2}\times\cdots$$

$$-\frac{1}{2}! = -\frac{1}{2}\left(-\frac{1}{2}-1\right)\left(-\frac{1}{2}-2\right)\left(-\frac{1}{2}-3\right)\cdots = -\frac{1}{2}\times -\frac{3}{2}\times -\frac{5}{2}\times -\frac{7}{2}\times\cdots$$

6.6 Introduction of Infinite Series

Infinite series are introduced so that we can learn how to deal with an n! that does not terminate. When we divide $1 - x^3$ by $1 - x$ the long division terminates, because the x^3 term drops down..

(11)

$$\begin{array}{r|l} & 1+x+x^2 \\ \hline 1-x & 1+0+0\ -x^3 \\ & \underline{1-x} \\ & \quad x \qquad -x^3 \\ & \quad \underline{x-x^2} \\ & \qquad x^2-x^3 \\ & \qquad \underline{x^2-x^3} \\ & \qquad\quad 0 \end{array}$$

When we divide 1 by 1–x the long division process does not terminate, because there is no x^k term to drop down. The quotient appears as

(12*a*) $1+x+x^2+x^3+....$ *ad infinitum*

(12*b*)

$$\begin{array}{r|l} & 1+x+x^2+x^3+\cdots \\ \hline 1-x & 1 \\ & \underline{1-x} \\ & x \\ & \underline{x-x^2} \\ & x^2 \\ & \underline{x^2-x^3} \\ & x^3 \\ & \underline{x^3-x^4} \\ & x^4 \end{array}$$

How do we interpret this *series* of terms so that the result makes sense? We know we can add a finite number of terms, so we do that.

(13*a*) $S(n)=1+x+x^2+x^3+....+x^{n-1}$

(13*b*) $S(n)-xS(n)=1-x^n$

(13*c*) $(1-x)S(n)=1-x^n \quad \Rightarrow \quad S(n)=\dfrac{1-x^n}{1-x}=\dfrac{1}{1-x}-\dfrac{x^n}{1-x}$

Convergent series A series of terms is convergent if the sum of the terms is finite as n increases to infinity (where infinity is an alias for "as large as we please"). An alternative to saying "the sum of the terms is finite as n increases to infinity" is the *limit* symbolism.

(14) *If* $x<1$ *and* $x>-1$, *then the magnitude of* x *is less than 1*

and $\lim\limits_{n\to\infty} x^n=0$

Taking the sum S(n) to the limit we get

(15) $\lim\limits_{n\to\infty} S(n)=\lim\limits_{n\to\infty}\left(\dfrac{1}{1-x}-\dfrac{x^n}{1-x}\right)=\dfrac{1}{1-x}$

In this sense we can use the infinite series to represent 1/(1–x).

(16) $\dfrac{1}{1-x}=1+x+x^2+x^3+x^4+.... \quad \rightarrow \quad |x|<1$

The series *converges* to 1/(1–x). We refer to the series as a *convergent series*. When the magnitude of x is greater than 1 the series is said to *diverge*. When x=1 the series does not exist. This is a very complex subject we have barely begun to explore.

6.7 The Binomial Theorem (any index)

Binomial Theorem

Let n be any number, positive or negative, integral or fractional.

Then the function $(1+x)^n$ is represented by the binomal series

$$1+nx+\frac{n(n-1)}{2!}x^2+\frac{n(n-1)(n-2)}{3!}x^3+....+\frac{n!}{k!(n-k)!}x^k+....$$

Notes

(1) When n is a positive integer the series terminates and the sum is $(1+x)^n$ for all x.

(2) When n is not a positive integer the series is an infinite series that represents $(1+x)^n$ when $-1<x<1$.

(3) Proof of the binomial theorem is outside the scope of this text, because knowledge of the general theory of infinite series is required. The theorem may be used with confidence provided however that the user remembers that the sum of *k* terms of the series *approaches* the value of $(1+x)^n$ as *k* increases. In other words we do not try to add up an infinite number of terms.

Examples

Let $n=\frac{1}{2}$

$$(1+x)^{\frac{1}{2}}=1+\frac{1}{2}x+\frac{\frac{1}{2}\left(\frac{1}{2}-1\right)}{2\cdot 1}x^2+\frac{\frac{1}{2}\left(\frac{1}{2}-1\right)\left(\frac{1}{2}-2\right)}{3\cdot 2\cdot 1}x^3+....$$

$$(1+x)^{\frac{1}{2}}=1+\tfrac{1}{2}x-\tfrac{1}{8}x^2+\tfrac{1}{16}x^3+....$$

Let $n=-1$ *and* $x\rightarrow -x$

$$(1-x)^{-1}=1+(-1)(-x)+\frac{-1(-1-1)}{2\cdot 1}(-x)^2+\frac{-1(-1-1)(-1-2)}{3\cdot 2\cdot 1}(-x)^3+....$$

$$\frac{1}{1-x}=1+x+x^2+x^3+....$$

Problems 6

604 Simplify

1. $\dfrac{5!}{3!}$ 2. $\dfrac{9!}{6!}$ 3. $\dfrac{6!\cdot 8!}{7!\cdot 9!}$ 4. $\dfrac{4!+5!}{3!\cdot 4!}$

5. $\dfrac{5!\cdot 6!}{9!-7!}$ 6. $\dfrac{(n-1)!}{n!}$ 7. $\dfrac{p!}{(p-2)!}$ 8. $\dfrac{2k!}{(2k)!}$

9. $\dfrac{n!}{(n-r)!}$ 10. $\dfrac{(n-k-1)!}{(n-k+1)!}$ 11. $\dfrac{(n+1)!-n!}{n!+(n-1)!}$ 12. $\dfrac{[(2n+1)!]^2}{(2n)!(2n+2)!}$

13. *Show that $n!$, $n>1$, is always an even number.*

14. *Show that* $\dfrac{n(n-1)(n-2)\cdots(n-r+1)}{r!}=\dfrac{n!}{r!(n-r)!}$

15. *Show that* $\dfrac{n!}{k!(n-k)!}+\dfrac{n!}{(k+1)!(n-k-1)!}=\dfrac{(n+1)!}{(k+1)!(n-k)!}$

605 Expand by the binomial theorem and simplify each term

1. $(x+y)^4$ 2. $(x-2y)^6$ 3. $(\frac{1}{2}x-3y^3)^3$ 4. $(x^{\frac{1}{2}}+y^{\frac{1}{2}})^5$

606 Find the first four terms of the binomial expansion and simplify each term.

1. $(x^{\frac{2}{3}}-\frac{1}{3}y^{-2})^{11}$ 2. $(x^{-3}+\frac{2}{3}y^{\frac{3}{2}})^{10}$ 3. $(x^4-2y^{-4})^{\frac{1}{4}}$ 4. $(8x^3+3y^2)^{\frac{2}{3}}$

607 Find the last four terms of the binomial expansion and simplify each term.

1. $(\frac{1}{3}x^2+y^{-2})^{10}$ 2. $(\frac{1}{2}x^{-1}-ay^{\frac{1}{2}})^{12}$ 3. $(\frac{2}{5}x^{\frac{2}{3}}+y^{\frac{3}{2}})^{11}$

608 Find and simplify the specified term

1. *5th term of* $(e^{2x}+e^{-2x})^{10}$
2. *5th term of* $(x^3+3y^3)^{\frac{4}{3}}$
3. x^7 *term of* $(\frac{1}{2}+x)^{13}$
4. $x^{\frac{11}{2}}$ *term of* $(\frac{1}{4}x^{-1}+x)^{\frac{1}{2}}$

609 Find numerical value correct to 5 significant figures.

1. $(1.01)^9$ 2. $(99)^4$ 3. $(103)^{\frac{1}{2}}$ 4. $(10)^{\frac{2}{3}}$ 5. $(1.03)^{-7}$

7 Exponential and Logarithmic Functions

Polynomials (Chapter 5) are a sum of one or more terms each of which is an *algebraic power function* x^n. The *variable x* is called the *base* and the *constant n* is called the *exponent*. For example:

(1) $x^5 - 4x^4 + x^3 - 13x^2 - 6x^1 + 6x^0$

Exponential function If, on the other hand, we have a function b^x in which the variable x appears as the exponent while positive real number constant b appears as the base, then we have the *exponential function* b^x.

(2) $y = b^x \rightarrow \quad b > 1 \;\; and \;\; real$

Base b can be any number real or complex. However base b is real in this Chapter. Complex base b is another subject.

The graph of the exponential function lies entirely above the x axis. Exponential function b^x is positive for every real value of x.

(3) $if \;\; x < 0 \rightarrow \quad 0 < b^x < 1,$

$if \;\; x = 0 \rightarrow \quad b^x = 1,$

$if \;\; x > 0 \rightarrow \quad b^x > 1$

Corresponding to each real value if x there is one and only one value of b^x. The exponential function is a single valued function.

Examples Several examples of this new symbolism follow.

exponential form	*logarithmic form*	*exponential form*	*logarithmic form*
$64 = 4^3$	$\log_4 64 = 3$	$\frac{1}{32} = 8^{-\frac{5}{3}}$	$\log_8 \frac{1}{32} = -\frac{5}{3}$
$27 = 9^{\frac{3}{2}}$	$\log_9 27 = \frac{3}{2}$	$1 = 7^0$	$\log_7 1 = 0$

The Base e Exponential Function Leonhard Euler (1707-1783) defined the base e exponential function when he proved that sum of a power series solution to a first order differential equation is his number e raised to the $-\alpha x$ power $e^{-\alpha x}$. He also proved the important property that the derivative of $e^{-\alpha x}$ is $-\alpha e^{-\alpha x}$ and calculated e=2.718281827....

Consider the equation of a parallel RC electric circuit.

$$0 = \frac{1}{R} v(t) + C \frac{dv(t)}{dt} \qquad \Rightarrow \qquad -\frac{1}{RC} dt = \frac{dv}{v}$$

Integrate both sides using x as a dummy variable.

$$-\frac{1}{RC} \int_0^t dx = \int_0^t \frac{dv(x)}{v(x)} \rightarrow -\frac{1}{RC}(t-0) = \ln v(t) - \ln v(0)$$

$$-\frac{1}{RC} t = \ln \frac{v(t)}{v(0)} \qquad \Rightarrow \qquad v(t) = v(0) e^{-\frac{t}{RC}}$$

Plots of e^{-x} *and* $1-e^{-x}$

(the maximum 1 values are scaled to 0.9 to clearly show the plots)

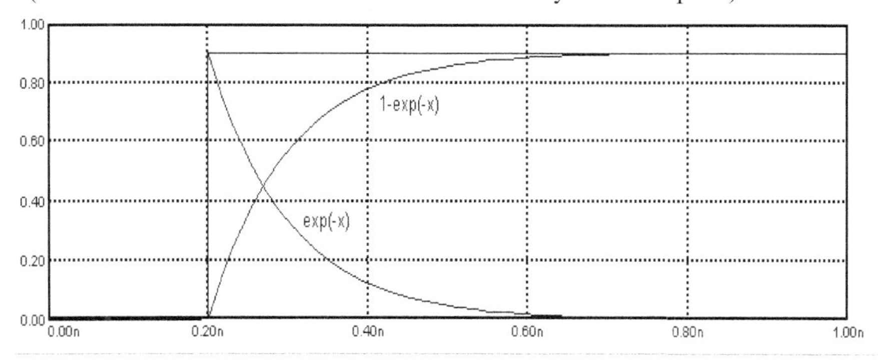

Problem 701 Let b=2. Plot $y=2^x$ when x = -3, -2, -1, 0, 1, 2, 3

Problem 702 Let b=10. Plot $y=10^x$ when x = -3, -2, -1, 0, 1, 2, 3

Problem 703 Let b= 1. Plot $y=1^x$ when x = -3, -2, -1, 0, 1, 2, 3.

Logarithmic function The logarithmic function is the inverse of the exponential function.

(4) $x = \log_b y \quad \Leftrightarrow \quad y = b^x \quad \rightarrow \quad b > 1$

Read this as "x is equal to the logarithm of y to the base b."

(5) *The exponent is* $x = \log_b y$ *and* $b^x = y$ *is the antilogarithm*

The base is b, the exponent or logarithm is x and the antilogarithm is y.

Bases most frequently used are 10 and e Although any positive number b greater than 1 may be taken as the base of a system of logarithms only two numbers are widely used in practice: 10 and Euler's e=2.71828…. , which appears in the Calculus. The base must be a positive number greater than 1. Why? If base b=1, then $y=b^x=1^x=1$, which goes nowhere.

The logarithm (log) *to the base 10* of a number is the exponent to which 10 must be raised to obtain the number.

Base 10 The log of 100 is 2, because $10^2=100$. The log of 0.001 is –3, because $10^{-3}=0.001$. The logarithms of most numbers are represented as decimals to so-many places. The log of 4.55 is 0.6580…., because $4.55=10^{0.6580}$.

Logarithms exist for all positive numbers, negative numbers and complex numbers. We only discuss the logarithm of positive numbers.

A discussion about calculating with logarithms conveniently starts with logarithms of numbers from 1 to 10 whose values range from 0 to 1. The range's 0, 1 end values are the 0 and 1 exponents of 10 ($10^0=1$, $10^1=10$).

The whole number part of a logarithm is referred to as the *characteristic*, and the decimal part is referred to as the *mantissa*.

For example, for log 4.55=0.6580 the 0 is the *characteristic* and .6580 is the *mantissa*.

Any number greater than 10 can be written as a number in the range 1 to 10 times a *positive* power of ten. We now show that the power of ten is the characteristic. For example when the power is 3 (see 7.1):

(6) $4550 = 4.55 \times 10^3 = 10^{0.6580} \times 10^3 = 10^{3.6580}$

and $\log 4550 = \log 10^{0.6580} + \log 10^3 = 0.6580 + 3 = 3.6580$

And any number from 0 to 1 can be written as a number in the range 1 to 10 times a *negative* power of ten.

(7) $0.0455 = 4.55\times10^{-2} = 10^{0.6580}\times10^{-2} = 10^{-1.3420}$ *and* $\log 0.0455 = -1.3420$

Gathering these results we can say
1 the logarithm of positive numbers greater than 1 is positive
2 the logarithm of 1 is zero
3 the logarithm of positive numbers less than 1 and >0 is negative.

Problem 704 Let b=2. Plot $\log_b 2^x$ when x = -3, -2, -1, 0, 1, 2, 3
Problem 705 Let b=10. Plot $\log_b 10^x$ when x = -3, -2, -1, 0, 1, 2, 3
Problem 706 Let b= 1. Plot $\log_b 1^x$ when x = -3, -2, -1, 0, 1, 2, 3.

7.1 Properties of Logarithms

Laws of exponents For example when x is an integer.

(8) $b^1 = b, \quad b^2 = b\times b, \quad b^3 = b\times b\times b$

Let b be any positive number, and x be any positive real number *not necessarily an integer*. Then the symbol b^x means the result of raising b to the power x. From this definition we have the following rules.

(9a) $b^x \cdot b^y = b^{x+y}$ *law for multiplication*

(9b) $\dfrac{b^x}{b^y} = b^{x-y}$ *law for division*

(9c) $(b^x)^y = b^{xy}$ *law for a power of a power*

The laws of exponents are valid for *all positive real numbers*.[1] (Not just integers.)

Negative Exponents

(10) *negative exponents* $y = b^{-x} = \dfrac{1}{b^x} = \dfrac{1}{m}$ *calculate* $m = b^x$, *then* $y = \dfrac{1}{m}$

For example

(11) $m = b^x = (0.7063)^{\frac{1}{6}} = 0.943695 \rightarrow y = \dfrac{1}{m} = 1.059665$

---

[1] The proof is beyond the scope of this text.

Properties of logarithms

1. The logarithm of a power q of a number w is equal to the power q times the logarithm of w, all logarithms being taken to the same base.

let	$p = \log_b w$
rewrite in exponential form	$w = b^p$
raise to the qth power	$w^q = (b^p)^q = b^{pq}$
rewrite in logarithmic form	$\log_b w^q = qp \log_b b = qp$
replace p by its log	$\log_b w^q = q \log_b w$

2. The logarithm of a product is equal to the sum of the logarithms of its factors, all logarithms being taken to the same base.

let	$p = \log_b w, \quad q = \log_b z$
rewrite in exponential form	$w = b^p, \ z = b^q$
multipy w by z	$wz = b^p b^q = b^{p+q}$
rewrite in logarithmic form	$\log_b wz = (p+q)\log_b b = p+q$
replace p and q by their values	$log_b wz = \log_b w + \log_b z$

3. The logarithm of a quotient is equal to the logarithm of the dividend minus the logarithm of the divisor, all logarithms being taken to the same base.

let	$p = \log_b w, \quad q = \log_b z$
rewrite in exponential form	$w = b^p, \ z = b^q$
divide w by z	$\frac{w}{z} = \frac{b^p}{b^q} = b^{p-q}$
rewrite in logarithmic form	$\log_b \frac{w}{z} = (p-q)\log_b b = p-q$
replace p and q by their logs	$\log_b \frac{w}{z} = \log_b w - \log_b z$

7.2 Solving Exponential and Logarithmic Equations

Exponential equations

(12) $2^{3x} = 4^{(x+1)}$

$3x\log 2 = (x+1)\log 4 = (x+1)\log 2^2 = 2(x+1)\log 2 \quad \textit{cancel } \log 2$

$3x = 2(x+1) = 2x+2 \quad \Rightarrow \quad x = 2$

For some problems you need a calculator (or a table of logs).

(13) $\mathrm{n} = 4.92^{5.368}$

$\log \mathrm{n} = 5.368 \times \log 4.92 = 5.368 \times 0.691965103 = 3.714468672$

$\log \mathrm{n} = 3 + 0.714468672 = \log 10^3 + \log 5.181657126 \quad \rightarrow \quad \mathrm{n} = 5.181657126 \times 10^3$

(14) $23.45 = b^{\frac{1}{7}}$

$\log 23.45 = \frac{1}{7}\log b \quad \Rightarrow \quad \log b = 7\log 23.45 = 7 \times 1.370143 = 9.59099993$

$\log b = 9 + 0.59099993 = \log 10^9 + \log 3.899424446 \quad \rightarrow \quad b = 3.899419229 \times 10^9$

(15) $801.2 = 14.56^d$

$\log 801.2 = d\log 14.56 \quad \Rightarrow \quad d = \frac{\log 801.2}{\log 14.56} = \frac{2.903740941}{1.163161375} = 2.496421394$

$d = 2.496421394$

Logarithmic equations Sometimes the numbers fall out.

(16) *solve for n* $\log_5 n = -2 \quad \rightarrow \quad n = 5^{-2} = \frac{1}{25}$

(17) *solve for b* $\log_b 16 = \frac{2}{3} \quad \rightarrow \quad 16 = b^{\frac{2}{3}} \quad \Rightarrow \quad b = 16^{\frac{3}{2}} = \left(16^{\frac{1}{2}}\right)^3 = 4^3 = 64$

(18) $2\log x - \log(30-2x) = 1 \quad \rightarrow \quad \log x^2 - \log(30-2x) = 1$

$$\log\frac{x^2}{30-2x} = 1$$

$$\frac{x^2}{30-2x} = 10^1 = 10$$

$$x^2 + 20x - 300 = 0 \quad \Rightarrow \quad x = -30, \quad x = 10$$

7.3 Logarithms to bases to other than 10

To find the relation between ln y (base e) and log y (base 10) start with

(20a) $y = \ln x \rightarrow e^y = x \rightarrow e^{\ln x} = x$

(20b) $y = \log x \rightarrow 10^y = x \rightarrow 10^{\log x} = x$

(21*a*) *from 20a we have* $p = e^{\ln p}$

(21*b*) $\log p = \log(e^{\ln p}) \quad \Rightarrow \quad \log p = \ln p \times \log(e)$

(21*c*) $\log e = \log 2.71828 = 0.43429$

(21*d*) $\log p = 0.43429 \;\ln p$

In general the relation between logarithms of the same number y to different bases b is given by the following theorem.

(22) Theorem $\quad \log_a y = \dfrac{\log_b y}{\log_b a}$

Proof let $x = \log_a y \rightarrow y = a^x$

$\rightarrow$ take base b log of both sides $\log_b y = x \log_b a \rightarrow x = \log_a y = \dfrac{\log_b y}{\log_b a}$

Useful Relations

(23) Let $y = b \quad \log_a b = \dfrac{\log_b b}{\log_b a} = \dfrac{1}{\log_b a}$

Important specific cases

(24) *Bases* $b = 10$ *and* $a = e \quad \log_e y = \dfrac{\log_{10} y}{\log_{10} e} \Rightarrow \ln y = \dfrac{\log y}{\log e}$

(25) *Bases* $a = 10$ *and* $b = e \quad \log_{10} y = \dfrac{\log_e y}{\log_e 10} \Rightarrow \log y = \dfrac{\ln y}{\ln 10}$

$\log e = 0.4343$ *and* $\ln 10 = 2.3026$ *and* $0.4343 = \dfrac{1}{2.3026}$

Problems 7

Convert exponential form y = b^x to logarithmic form x = $\log_b$ y.

1. $3^4 = 81$ 2. $8^{\frac{1}{3}} = 2$ 3. $2^3 = 8$ 4. $10^4 = 10000$ 5. $10^{-2} = \frac{1}{100}$

6. $\left(\frac{1}{4}\right)^3 = \frac{1}{64}$ 7. $7^x = y$ 8. $b^3 = y$ 9. $10^x = y$ 10. $b^x = 27$

Convert logarithmic form x = $\log_b$ y to exponential form y = b^x.

11. $5 = \log_2 32$ 12. $\frac{1}{4} = \log_{16} 2$ 13. $4 = \log_2 16$

14. $6 = \log_{10} 1000000$ 15. $-3 = \log_2 \frac{1}{8}$

16. $-3 = -\log_2 8$ 17. $z = \log_{10} y$ 18. $5 = \log_3 243$

19. $-3 = \log_{10} 0.001$ 20. $x = \log_3 81$

Find $\log_{10}$ of the following numbers.

21. 100 22. 0.01 23. 1000 24. 1 25. 0.001

26. 100000 27. 0.00001 28. 10 29. 0.1 30. 0.001

Solve for x.

31. $\log_{10} x = 5$ 32. $\log_x 16 = 4$ 33. $\log_2 x = 5$ 34. $\log_4 64 = x$

35. $\log_{16} x = \frac{3}{2}$ 36. $\log_x 27 = \frac{3}{4}$ 37. $\log_{25} 625 = x$ 38. $\log_4 x = \frac{5}{2}$

Find characteristic of $\log_{10}$ y for following y.

39. 7.234 40. 72.34 41. 0.7234 42. 72340 43. 7234×10^4

44. 0.007234 45. 72.34×10^{-6} 46. 723400 47. 0.72340×10^{-4}

If $\log_{10}$ y = 0.69897, then y = 5. Use laws for log xy and log x/y to find log of the following numbers.

48. $\log 10y$ 49. $\log \frac{y}{10}$ 50. $\log 100y$ 51. $\log 1000y$ 52. $\log \frac{y}{1000}$

53. $\log 2(\textit{hint } 2 = \frac{10}{5})$ 54. $\log \frac{2}{10}$ 55. $\log 200$ 56. $\log \frac{1}{2}$ 57. $\log \frac{100}{2}$

Log 10 = 1, Log 5 = 0.69897, 1–0.69897 = 0.30103. Find the antilog of the following numbers.

58. $1-0.69897$ 59. 2.69897 60. $-1+0.30103$ 61. $2+3.69897$

62. $-1+0.69897$ 63. 4.30103 64. 2.30103–1.69897 65. 2.69897–1.30103

Find the value of these expressions.

66. $\log 10^3$ 67. $\log(0.01)^4$ 68. $\log(0.001)^3$ 69. $\log 5^3$ 70. $\log 2^4$

Expand as algebraic sum of terms.

71. $\log 3^2 7^3 5^7$ 72. $\log 9^{-1} 7^2$ 73. $\log 4^{\frac{1}{2}} 8^{\frac{1}{3}}$ 74. $\log 5^2 4^3$ 75. $3\log 5^2 7$

8 Partial Fractions

In many problems a *rational function*, the ratio of two polynomials N(p)/D(p), is decomposed into a sum of fractions with denominators of lower degree. Each fraction in the sum is referred to as a *partial fraction*. This process is the inverse of a process that adds fractions.

Given a ratio of two polynomials in the variable p, we can always divide the denominator into the numerator so that the remainder F(p) is a proper rational fraction. This means the degree of N(p) is less than the degree of D(p). For example:

$$(1)\quad if\;\; G(p)=\frac{p^4+5p^3+8p^2+3p+6}{p^3+4p^2+3p}=p+1+\frac{p^2+6}{p^3+4p^2+3p}$$

$$(2)\quad let\; F(p)=\frac{p^2+6}{p^3+4p^2+3p}=\frac{N(p)}{D(p)}\quad (a\;\; proper\;\; fraction)$$

There are two problems to solve:
1. Find the roots of D(p) (Chapter 4).

2. Expand F(p) into a sum of partial fractions.

The basis for decomposing a proper fraction into a sum of partial fractions is the *Theorem* in the side bar, the proof[1] of which is beyond the scope of this text.

The methods decomposing a ratio of two polynomials N(p)/D(p) into a sum of fractions are based on the fact coefficients of similar terms of two equal polynomials are equal.

---

[1] W. L. Ferrar *Higher Algebra* Oxford Press ISBN 0198325061

Theorem *Any proper fraction N(p)/D(p) whose numerator and denominator are polynomials in p may be decomposed into an algebraic sum of partial fractions of the types listed here.*

1. If any linear factor, such as (ap+b), occurs *once* as a factor of the denominator D(p) of the given function there will correspond to that factor a partial fraction with the form

$$\frac{A}{ap+b}$$

2. If any linear factor, such as (ap+b), occurs *k* times as a factor of the denominator of D(p) of the given function there will correspond to that factor the sum of *k* partial fractions with the form

$$\frac{A_1}{ap+b}+\frac{A_2}{(ap+b)^2}+\cdots+\frac{A_k}{(ap+b)^k} \quad \text{where the A's are constants and } A_k \neq 0$$

3. If any quadratic factor, such as (ap^2+bp+c), occurs *once* as a factor of the denominator of D(p) of the given function there will correspond to that factor a partial fraction with the form

$$\frac{Ap+B}{ap^2+bp+c}$$

4. If any quadratic factor, such as (ap^2+bp+c), occurs *k* times as a factor of the denominator of D(p) of the given function there will correspond to that factor the sum of *k* partial fractions with the form

$$\frac{A_1p+B_1}{ap^2+bp+c}+\frac{A_2p+B_2}{(ap^2+bp+c)^2}+\cdots+\frac{A_kp+B_k}{(ap^2+bp+c)^k}$$

Note: In every case the number of constants in the numerator of a partial fraction equals the degree n of the denominator D(p) of the fraction. The numerator is then a polynomial of degree n–1.

8.1 Sums of Fractions

We know from arithmetic that it is always possible to express the sum or difference of a number of fractions as one fraction. The same can be said for fractions with polynomials. Consider these examples of proper fraction sums implemented by using the trick a/a=1.

create common denominator by multiplying by a/a and b/b

$$(3)\quad f(p)=\frac{1}{p+1}-\frac{1}{p+2}=\frac{1}{p+1}\frac{a}{a}-\frac{1}{p+2}\frac{b}{b}\quad\rightarrow\quad a=p+2\ \ and\ \ b=p+1$$

$$=\frac{1}{p+1}\cdot\frac{p+2}{p+2}-\frac{1}{p+2}\cdot\frac{p+1}{p+1}$$

fraction manipulations :

$$\tfrac{n_1}{d}-\tfrac{n_2}{d}=\tfrac{1}{d}\tfrac{n_1}{1}-\tfrac{1}{d}\tfrac{n_2}{1}=\tfrac{1}{d}\left(\tfrac{n_1}{1}-\tfrac{n_2}{1}\right)=\tfrac{1}{d}\left(n_1-n_2\right)=\tfrac{1}{d}\left(\tfrac{n_1-n_2}{1}\right)=\tfrac{n_1-n_2}{d}$$

$$f(p)=\frac{(p+2)-(p+1)}{(p+1)(p+2)}=\frac{p+2-p-1}{(p+1)(p+2)}=\frac{1}{(p+1)(p+2)}$$

$$(4)\quad f(p)=\frac{1}{p+1}-\frac{2}{p+2}=\frac{(p+2)-2(p+1)}{(p+1)(p+2)}=-\frac{p}{(p+1)(p+2)}$$

$$(5)\quad f(p)=\frac{p+1}{p^2+1}+\frac{2p}{p^2+2}=\frac{(p+1)(p^2+2)+2p(p^2+1)}{(p^2+1)(p^2+2)}$$

$$=\frac{p(p^2+2)+1(p^2+2)+2p(p^2+1)}{(p^2+1)(p^2+2)}$$

$$=\frac{p^3+2p+p^2+2+2p^3+2p}{(p^2+1)(p^2+2)}=\frac{3p^3+p^2+4p+2}{(p^2+1)(p^2+2)}$$

Emphasis: Observe that the sums are also proper fractions. The degree of the numerator is less than the degree of the denominator.

In many branches of mathematics the reverse operation needs to be executed. Given a fraction whose denominator is factored, express the fraction as an algebraic sum of partial fractions, where each partial fraction is a proper fraction.

8.2 Linear Factors of order 1

When the denominator of a proper fraction F(p) can be resolved into real linear factors, all of which are distinct.

Consequently each partial fraction has to be a proper fraction, where A_1, A_2, and A_3 are real constants so that the degree of the numerators less than the degree of the denominators. Reference Theorem part 1.

$$(6) \quad F(p) = \frac{p^2+6}{p^3+4p^2+3p}$$

$$(7) \quad \frac{p^2+6}{p(p+1)(p+3)} = \frac{A_1}{p} + \frac{A_2}{p+1} + \frac{A_3}{p+3}$$

cross multiply :

$$(8) \quad p^2+6 = A_1(p+1)(p+3) + A_2 p(p+3) + A_3 p(p+1)$$

Do not simplify by multiplying out. It is easier to substitute selected values of p.

(9) *if* $p=0$ *then*

$$0+6 = A_1(0+1)(0+3) + A_2 0(0+3) + A_3 0(0+1)$$

$$6 = 3A_1 \quad \Rightarrow \quad A_1 = 2$$

(10) *if* $p=-1$ *then*

$$1+6 = A_1(-1+1)(-1+3) + A_2(-1)(-1+3) + A_3(-1)(-1+1)$$

$$7 = -2A_2 \quad \Rightarrow \quad A_2 = -\frac{7}{2}$$

(11) *if* $p=-3$ *then*

$$9+6 = A_1(-3+1)(-3+3) + A_2(-3)(-3+3) + A_3(-3)(-3+1)$$

$$15 = 6A_3 \quad \Rightarrow \quad A_3 = \frac{5}{2}$$

Check

$$(12) \quad F(p) = \frac{A_1}{p} + \frac{A_2}{p+1} + \frac{A_3}{p+3} = \frac{2}{p} + \frac{-7/2}{p+1} + \frac{5/2}{p+3}$$

$$2F(p) = \frac{4}{p} - \frac{7}{p+1} + \frac{5}{p+3} = \frac{4(p+1)(p+3) - 7p(p+3) + 5p(p+1)}{p(p+1)(p+3)}$$

$$F(p) = \frac{1}{2}\frac{(4-7+5)p^2 + (16-21+5)p + 12}{p(p+1)(p+3)} = \frac{p^2+6}{p(p+1)(p+3)} \quad qed$$

8.3 Linear Factors of order k

When the denominator of a proper fraction F(p) can be resolved into real linear factors, some of which are repeated.

As before each partial fraction has to be a proper fraction.

Each higher order factor of order n requires a sum of terms. There is one term for each power of p, or p+1, from 1 to k. Reference Theorem part 2.

$$(13)\quad F(p) = \frac{5p^3 - 6p - 3}{p^3(p+1)^2} = \frac{A_1}{p} + \frac{A_2}{p^2} + \frac{A_3}{p^3} + \frac{A_4}{p+1} + \frac{A_5}{(p+1)^2}$$

cross multiply

$$(14a)\quad 5p^3 - 6p - 3 = A_1p^2(p+1)^2 + A_2p(p+1)^2 + A_3(p+1)^2 + A_4p^3(p+1) + A_5p^3$$

$$(14b)\quad 5p^3 - 6p - 3 = A_1(p^4 + 2p^3 + p^2) + A_2(p^3 + 2p^2 + p) + A_3(p^2 + 2p + 1) + A_4(p^4 + p^3) + A_5p^3$$

$$(15)\quad 5p^3 - 6p - 3 = p^4(A_1 + A_4) + p^3(2A_1 + A_2 + A_4 + A_5) + p^2(A_1 + 2A_2 + A_3) + p(A_2 + 2A_3) + 1(A_3)$$

From 14a

Let $p = 0 \qquad -3 = 0+0+0+A_3+0+0 \qquad \Rightarrow \quad A_3 = -3$

Let $p = -1 \qquad -5+6-3 = 0+0+0+0+0-A_5 \qquad \Rightarrow \quad A_5 = 2$

From 15

equate coefficients of terms

$p^4 : 0 = A_1 + A_4 \qquad p^3 : 5 = 2A_1 + A_2 + A_4 + A_5 \qquad p^2 : 0 = A_1 + 2A_2 + A_3$

$p : -6 = A_2 + 2A_3 \qquad 1 : -3 = A_3$

$$(16)\quad A_3 = -3,\ A_2 = 0,\ A_1 = 3,\ A_4 = -3,\ A_5 = 2$$

$$(17)\quad F(p) = \frac{5p^3 - 6p - 3}{p^3(p+1)^2} = \frac{3}{p} + \frac{0}{p^2} - \frac{3}{p^3} - \frac{3}{p+1} + \frac{2}{(p+1)^2}$$

Check

$$(18)\ F(p)=\frac{A_1}{p}+\frac{A_2}{p^2}+\frac{A_3}{p^3}+\frac{A_4}{p+1}+\frac{A_5}{(p+1)^2}=\frac{3}{p}+\frac{0}{p^2}-\frac{3}{p^3}-\frac{3}{p+1}+\frac{2}{(p+1)^2}$$

$$=\frac{3p^2(p+1)^2-3(p+1)^2-3p^3(p+1)+2p^3}{p^3(p+1)^2}$$

$$=\frac{3(p+1)[p^2(p+1)-(p+1)-p^3]+2p^3}{p^3(p+1)^2}=\frac{3(p+1)[p^3+p^2-p-1-p^3]+2p^3}{p^3(p+1)^2}$$

$$=\frac{(3p+3)[p^2-p-1]+2p^3}{p^3(p+1)^2}=\frac{[3p^3+3p^2-3p^2-3p-3p-3]+2p^3}{p^3(p+1)^2}$$

$$=\frac{5p^3-6p-3}{p^3(p+1)^2}\quad qed$$

8.4 Quadratic Factors of order 1

When the denominator of a proper fraction F(p) contains one quadratic factor, but no repeated quadratic factor. Reference Theorem part 3.

$$(19)\quad F(p)=\frac{16}{p(p^2+5p+2)}$$

$$(20)\quad \frac{16}{p(p^2+5p+2)}=\frac{A_1}{p}+\frac{A_2p+B_2}{p^2+5p+2}\qquad \textit{now cross multiply}$$

$$16=A_1(p^2+5p+2)+(A_2p+B_2)p$$

$$(21)\quad 16=A_1p^2+5A_1p+2A_1+A_2p^2+B_2p$$

equate coefficients of terms

$p^2:\ 0=A_1+A_2 \qquad p:\ 0=5A_1+B_2 \qquad 1:\ 16=2A_1$

$$(22)\quad A_1=8,\ A_2=-8,\ B_2=-40$$

$$(23)\quad F(p)=\frac{16}{p(p^2+5p+2)}=\frac{8}{p}-\frac{8p+40}{p^2+5p+2}$$

Check

$$(24)\quad F(p)=\frac{A_1}{p}+\frac{A_2p+B_2}{p^2+5p+2}=\frac{8}{p}-\frac{8p+40}{p^2+5p+2}=\frac{8(p^2+5p+2)-p(8p+40)}{p(p^2+5p+2)}$$

$$=\frac{8p^2+40p+16-8p^2-40p}{p(p^2+5p+2)}=\frac{16}{p(p^2+5p+2)}\quad qed$$

8.5 Quadratic Factors of order k

When the denominator of a proper fraction F(p) contains one quadratic repeated factor. Reference Theorem part 4.

$$(25)\quad F(p)=\frac{16}{p(p^2+5p+2)^2}$$

$$(26)\quad \frac{16}{p(p^2+5p+2)^2}=\frac{A_1}{p}+\frac{A_2p+B_2}{p^2+5p+2}+\frac{A_3p+B_3}{(p^2+5p+2)^2}$$

Now cross multiply.

$$16=A_1(p^2+5p+2)^2+(A_2p+B_2)p(p^2+5p+2)+(A_3p+B_3)p$$

$$(27)\quad 16=A_1(p^4+10p^3+29p^2+20p+4)$$
$$+A_2(p^4+5p^3+2p^2)+B_2(p^3+5p^2+2p)+A_3p^2+B_3p$$
$$=p^4(A_1+A_2)+p^3(10A_1+5A_2+B_2)$$
$$+p^2(29A_1+2A_2+5B_2+A_3)+p(20A_1+2B_2+B_3)+1(4A_1)$$

equate coefficients of terms

$p^4:\ 0=A_1+A_2$ $\qquad p^3:\ 0=10A_1+5A_2+B_2$

$p^2:\ 0=29A_1+2A_2+5B_2+A_3$ $\qquad p:\ 0=20A_1+2B_2+B_3$ $\qquad 1:\ 16=4A_1$

$$(28)\quad A_1=4,\ A_2=-4,\ B_2=-20,\ B_3=-40,\ A_3=-8$$

Check

$$(29)\quad F(p)=\frac{A_1}{p}+\frac{A_2p+B_2}{p^2+5p+2}+\frac{A_3p+B_3}{(p^2+5p+2)^2}$$
$$=\frac{4}{p}-\frac{4p+20}{p^2+5p+2}-\frac{8p+40}{(p^2+5p+2)^2}$$
$$=\frac{4(p^2+5p+2)^2-p(4p+20)(p^2+5p+2)-p(8p-40)}{p(p^2+5p+2)^2}$$
$$=\frac{(p^2+5p+2)(4p^2+20p+8-4p^2-20p)-8p^2-40p}{p(p^2+5p+2)^2}$$
$$=\frac{(p^2+5p+2)(8)-8p^2-40p}{p(p^2+5p+2)^2}=\frac{16}{p(p^2+5p+2)^2}\quad qed$$

Problems 8

Convert into a sum of partial fractions as shown. Show all steps

1 $\dfrac{p-13}{(p-3)(p+2)}=\dfrac{3}{p+2}-\dfrac{2}{p-3}$

2 $\dfrac{x+2}{2x^2-x}=\dfrac{5}{2x-1}-\dfrac{2}{x}$

3 $\dfrac{3y^2+4y-15}{(y-3)(y-1)(y+1)}=\dfrac{3}{y-3}+\dfrac{2}{y-1}-\dfrac{2}{y+1}$

4 $\dfrac{z^2-2z+16}{z^3-4z}=-\dfrac{4}{z}+\dfrac{2}{z-2}+\dfrac{3}{z+2}$

5 $\dfrac{x^2-11x-6}{2x^3-x^2-8x+4}=\dfrac{3}{2x-1}-\dfrac{2}{x-2}+\dfrac{1}{x+2}$

6 $\dfrac{x^2}{(x+1)^2(x-1)}=\dfrac{1}{4(x-1)}+\dfrac{3}{4(x+1)}-\dfrac{1}{2(x+1)^2}$

7 $\dfrac{15-12x}{(x-2)^2(2x-1)^2}=\dfrac{4}{(2x-1)^2}-\dfrac{1}{(x-2)^2}$

8 $\dfrac{x^2-x+13}{(x+1)(x^2+4)}=\dfrac{3}{x+1}+\dfrac{1-2x}{x^2+4}$

9 $\dfrac{x^4-2x^3+2x^2-2x+2}{x^3-x^2+x-1}=x-1+\dfrac{1}{2}\left(\dfrac{1}{(x-1)}-\dfrac{x+1}{(x^2+1)}\right)$

9 Mathematical Induction

The mathematical induction method of *proof by induction* has many uses such as proving theorems, discovering new results, and providing relatively simple proofs of theorems obtained by other means.

The Principal of Mathematical induction A mathematical formula involving the positive integer n is true for all positive values of n provided that (1) the formula is true when n=1, and (2) the hypothesis that the formula is true for any n is sufficient to ensure that the formula is true for n+1.

Suppose a formula satisfies conditions (1) and (2). Then by (2) if the formula is true for n, it is also true for n+1.

Therefore if true for n=1 it is true for n=2,
if true for n=2 it is true for n=3,
and so for all values of n.

Example 1 Prove by induction that (1) $1^3+2^3+3^3+....+n^3=\frac{1}{4}n^2(n+1)^2$

$(2a)\ \ n=1 \ \rightarrow\ 1^3=1\ \ and\ \ \frac{1}{4}1^2(1+1)^2=1$

$(2b)\ \ n=2 \ \rightarrow\ 1^3+2^3=9\ \ and\ \ \frac{1}{4}2^2(2+1)^2=9$

$(2c)\ \ n=n \ \rightarrow\ 1^3+2^3+3^3+....+n^3=\frac{1}{4}n^2(n+1)^2$

$(2d)\ \ n=n+1 \ \rightarrow\ add\ (n+1)^3\ to\ both\ sides\ of\ =$

$1^3+2^3+3^3+....+n^3+(n+1)^3=\frac{1}{4}n^2(n+1)^2+(n+1)^3$

$=\frac{1}{4}(n+1)^2[n^2+4(n+1)]=\frac{1}{4}(n+1)^2[n^2+4n+4)]=\frac{1}{4}(n+1)^2[(n+2)^2]\ \ qed$

Example 2 Prove by induction that (3) $1^2+2^2+....+n^2=\frac{1}{6}n(n+1)(2n+1)$

$(4a)\ \ n=1 \ \rightarrow\ 1^2=1\ \ and\ \ \frac{1}{6}1(1+1)(2+1)=1$

$(4b)\ \ n=2 \ \rightarrow\ 1^2+2^2=5\ \ and\ \ \frac{1}{6}2(2+1)(4+1)=5$

$(4c)\ \ n=n+1 \ \rightarrow\ add\ (n+1)^2\ to\ both\ sides\ of\ =$

$1^2+2^2+3^2+....+n^2+(n+1)^2=\frac{1}{6}n(n+1)(2n+1)+(n+1)^2$

$=\frac{1}{6}n(n+1)(2n+1)+(n+1)^2$

$=\frac{1}{6}(n+1)[n(2n+1)+6(n+1)]=\frac{1}{6}(n+1)[2n^2+7n+6]=\frac{1}{6}(n+1)[(2n+3)(n+2)]\ \ qed$

10 Progressions

The two basic types of progressions are arithmetic and geometric.

An arithmetic progression (AP) is a sequence of numbers in which each term, after the first, is obtained from the preceding one by adding to it a fixed number referred to as the common difference,

A geometric progression (GP) is a sequence of numbers in which each term, after the first, is obtained from the preceding one by multiplying it by a fixed number referred to as the common ratio,

10.1 Arithmetic Progressions

If x is the first term of an AP then x+d is the next term where d is the common difference. Then the nth term is x+(n–1)d. For example:

$$\text{If } x = 3 \text{ and } d = 5 \text{ then } S = 3 + 8 + 13 + 18 + \ldots$$

$$\text{If } x = \frac{7}{12} \text{ and } d = -\frac{1}{12} \text{ then } S = \frac{7}{12} + \frac{6}{12} + \frac{5}{12} + \frac{4}{12} + \ldots$$

$$\text{If } x = a \text{ and } d = b - a \text{ then}$$

$$S = a + [a + (b - a)] + [a + 2(b - a)] + [a + 3(b - a)] + \ldots$$

$$S = a + b + (2b - a) + (3b - 2a) + \ldots$$

The sum of the first n terms of an AP is readily found. For example:

$$S_4 = x \qquad + (x + d) \quad + (x + 2d) + (x + 3d)$$

$$S_4 = (x + 3d) + (x + 2d) + (x + d) \quad + x$$

$$2S_4 = (2x + 3d) + (2x + 3d) + (2x + 3d) + (2x + 3d) = 4(2x + 3d)$$

$$S_4 = 2(2x + 3d)$$

$$\text{And for n terms} \quad 2 \to n/2 \qquad 3 \to n - 1$$

$$S_n = \tfrac{1}{2}n[2x + (n - 1)d)$$

The terms of an AP between any two terms are the Arithmetic Means between those two terms. If only one term is between any two terms, then the term is the Arithmetic Mean of the two terms.

10.2 Geometric Progressions

If x is the first term of a GP then rx is the next term where r is the common ratio. Then the nth term is xr^{n-1}. For example:

If $x = 7$ and $r = 2$ then $S = 7 + 7\times 2 + 7\times 2^2 + 7\times 2^3 + \ldots$

If $x = \frac{7}{12}$ and $r = -\frac{1}{12}$ then $S = \frac{7}{12} + \frac{7}{12}\left(-\frac{1}{12}\right) + \frac{7}{12}\left(-\frac{1}{12}\right)^2 + \frac{7}{12}\left(-\frac{1}{12}\right)^3 + \ldots$

If $x = a$ and $r = b - a$ then $S = a + a(b-a) + a(b-a)^2 + a(b-a)^3 + \ldots$

The sum of the first n terms of a GP is promptly found. For example:

$$S_4 = x + xr + xr^2 + xr^3$$

$$rS_4 = xr + xr^2 + xr^3 + xr^4$$

$$S_4 - rS_4 = x - xr^4$$

$$S_4 = x\frac{1-r^4}{1-r} \quad r \neq 1$$

And for n terms $S_n = x\frac{1-r^n}{1-r} \quad r \neq 1$

$$S_n = \frac{x}{1-r} - \frac{xr^n}{1-r} \quad r < 1$$

$$\lim_{n\to\infty} r^n = 0 \text{ so that } \lim_{n\to\infty} S_n = \frac{x}{1-r}$$

The terms of a GP between any two terms are the Geometric Means between those two terms. If only one term is between any two terms, then the term is the Geometric Mean of the two terms.

Repeating decimal to fraction conversion For example;

$$d = 0.3272727\cdots = 0.3 + 0.027 + 0.00027 + 0.0000027 + \ldots$$

$$x = 0.027 \text{ and } r = 0.01 = 10^{-2}$$

$$\lim_{n\to\infty} S_n = \frac{x}{1-r} = \frac{0.027}{1-0.01} = \frac{0.027}{0.99} = \frac{27}{990} = \frac{3\times 9}{110\times 9} = \frac{3}{110}$$

$$d = 0.3 + \frac{3}{110} = \frac{3}{10}\frac{11}{11} + \frac{3}{110} = \frac{36}{110} = \frac{18}{55}$$

11 Inequalities

An inequality is a statement that one algebraic expression is greater than or less than another algebraic expression. Each expression is a member of the inequality and must be a real number here. For example;

$x > y$ x is greater than y and $x - y = n$ is a positive number

$x < y$ x is less than y and $x - y = -n$ is a negative number

$x \geq y$ x is greater than or equal to y

$x \leq y$ x is less than or equal to y

Sense: Two inequalities are said to be of the same sense if their symbols point in the same direction. $a > b$ and $c > d$ have the same sense.

Absolute inequality: If the sense of an inequality is the same for all values of the symbols for which its members are defined, the inequality is an absolute or unconditional inequality. For example:

$x^2 + y^2 > 0$ if either $x \neq 0$ or $y \neq 0$

$9 > 6$

$\pi < 22/7$

Conditional inequality: If the sense of an inequality holds for only for certain values of the symbols involved, but is not true for other values, the inequality is a conditional inequality.

$x + 2 < 5$ true only when $x < 3$

$x^2 + 4 > 5x$ true only when $x < 1$ or $x > 4$

Properties of Inequalities: Proof of property 1 is given. Proofs of the remaining properties are left as problems to be solved.

1 The sense of an inequality is not changed if both members are increased or decreased by the same number/

$a > b$ by hypothesis

$\therefore\ a - b = n$

$a + c - b - c = n$

$(a + c) - (b + c) = n$

$a + c > b + c$ qed

2 The sense of an inequality is not changed if both members are multiplied or divided by the same positive number/

if $x > y$ then $ax > ay$ and $x/c > y/c$ when $c > 0$

3 The sense of an inequality is reversed if both members are multiplied or divided by the same negative number/

if $x > y$ then $ax < ay$ when $a < 0$ and $x/c < y/c$ when $c < 0$

4 The sense of an inequality whose members are positive numbers is not changed by taking like positive powers or roots of both members.

if $x > y$ then $x^m > y^m$ and $x^{1/n} > y^{1/n}$ where x, y, m, n are > 0

5 If positive number is divided by members of an inequality which are of like sign, the resulting inequality is of the opposite sense.

if $c > 0$ and $x > y$ then $(c \div x) < (c \div y)$ when x and y have like signs and $c > 0$
if $c = 7$ and $5 > 4$ then $7/5 < 7/4$

Examples:

Show that $\frac{x}{y} + \frac{y}{x} > 2$ if $x \neq y$

$$\frac{x^2 + y^2}{xy} > 2 \quad \rightarrow \quad x^2 + y^2 > 2xy$$

$$x^2 + y^2 - 2xy > 0 \quad \rightarrow \quad (x - y)^2 > 0 \quad \text{or} \quad (y - x)^2 > 0 \quad \text{qed}$$

Show that $\frac{x - y}{x + y} < \frac{x^2 - y^2}{x^2 + y^2}$ if $x > y$

$$(x - y)(x^2 + y^2) < (x^2 - y^2)(x + y)$$

$$x^3 + xy^2 - yx^2 - y^3 < x^3 + yx^2 - y^2x - y^3$$

$$2xy^2 < 2yx^2 \quad \rightarrow \quad y^2 < yx \quad \rightarrow \quad y < x \quad \text{qed}$$

12 Matrix Algebra

A matrix is an array of r×c numbers, real or complex, arranged in r rows and c columns. Matrices allow one to write and process equations efficiently. Furthermore, in many problems, the matrix format makes the next step easier to perceive. This will become clear as we proceed. The world says a matrix has r rows and c columns. This matrix has 2 rows and 3 columns. D is a 2×3 matrix.

$$(1)\quad D = D_{row\times column} = D_{2\times 3} = \begin{bmatrix} 3 & 7 & 9 \\ 4 & 5 & -1 \end{bmatrix}$$

$$(2a)\quad rows \qquad (3\ \ 7\ \ 9)\quad (4\ \ 5\ \ -1)$$

$$(2b)\quad columns \qquad \begin{pmatrix} 3 \\ 4 \end{pmatrix}\ \begin{pmatrix} 7 \\ 5 \end{pmatrix}\ \begin{pmatrix} 9 \\ -1 \end{pmatrix}$$

Matrix equation AX=B represents two equations.

$$(3a)\quad 2x + 9y = -7 \qquad (2b)\quad 5x + 3y = 1$$

$$(3b)\quad \begin{bmatrix} 2 & 9 \\ 5 & 3 \end{bmatrix} \times \begin{bmatrix} x \\ y \end{bmatrix} = \begin{bmatrix} -7 \\ 1 \end{bmatrix}$$

$$(3c)\quad AX = B$$

The zero, null, 2×2 matrix is $(4)\ \begin{bmatrix} 0 & 0 \\ 0 & 0 \end{bmatrix}$

12.1 Matrix Addition and Subtraction

Add matrices by adding corresponding elements. Subtract by replacing + by –. Observe that addition and subtraction requires *conformable* matrices. For example consider 2×3 and 3×2 matrices.

$$(5a)\quad A + B = \begin{bmatrix} a_{11} & a_{12} & a_{13} \\ a_{21} & a_{22} & a_{23} \end{bmatrix} + \begin{bmatrix} b_{11} & b_{12} & b_{13} \\ b_{21} & b_{22} & b_{23} \end{bmatrix} = \begin{bmatrix} a_{11}+b_{11} & a_{12}+b_{12} & a_{13}+b_{13} \\ a_{21}+b_{21} & a_{22}+b_{22} & a_{23}+b_{23} \end{bmatrix}$$

$$(5b)\quad C + D = \begin{bmatrix} a_{11} & a_{12} \\ a_{21} & a_{22} \\ a_{31} & a_{32} \end{bmatrix} + \begin{bmatrix} b_{11} & b_{12} \\ b_{21} & b_{22} \\ b_{31} & b_{32} \end{bmatrix} = \begin{bmatrix} a_{11}+b_{11} & a_{12}+b_{12} \\ a_{21}+b_{21} & a_{22}+b_{22} \\ a_{31}+b_{31} & a_{32}+b_{32} \end{bmatrix}$$

Two matrices are conformable for addition when each has the same number of rows, and each has the same number of columns.

12.2 Matrix Multiplication

Any number q times matrix A multiplies each element of A q times. Let q=3.

$$(6a)\quad 3A = A + A + A = \begin{bmatrix} a_{11} & a_{12} \\ a_{21} & a_{22} \end{bmatrix} + \begin{bmatrix} a_{11} & a_{12} \\ a_{21} & a_{22} \end{bmatrix} + \begin{bmatrix} a_{11} & a_{12} \\ a_{21} & a_{22} \end{bmatrix} = \begin{bmatrix} 3a_{11} & 3a_{12} \\ 3a_{21} & 3a_{22} \end{bmatrix}$$

$$(6b)\quad qA = \begin{bmatrix} qa_{11} & qa_{12} \\ qa_{21} & qa_{22} \end{bmatrix}$$

The matrix –B is a matrix whose elements are those of B multiplied by –1. I.e. q = –1. We can demonstrate this by subtraction 0–B = –B.

$$(7a)\quad 0 - B = \begin{bmatrix} 0 & 0 \\ 0 & 0 \end{bmatrix} - \begin{bmatrix} b_{11} & b_{12} \\ b_{21} & b_{22} \end{bmatrix} = \begin{bmatrix} 0 - b_{11} & 0 - b_{12} \\ 0 - b_{21} & 0 - b_{22} \end{bmatrix} = \begin{bmatrix} -b_{11} & -b_{12} \\ -b_{21} & -b_{22} \end{bmatrix} = -B$$

$$(7b)\quad \text{If } B = \begin{bmatrix} b_{11} & b_{12} \\ b_{21} & b_{22} \end{bmatrix} \text{ then } -B = -1 \times B = \begin{bmatrix} -b_{11} & -b_{12} \\ -b_{21} & -b_{22} \end{bmatrix}$$

The dot, scalar, or inner product of two numbers is a guide to matrix multiplication.

$$(8a)\quad x \text{ has components } x_1\ x_2\ x_3 \ldots x_n$$

$$(8b)\quad y \text{ has components } y_1\ y_2\ y_3 \ldots y_n$$

$$(8c)\quad \text{then the dot product } x \cdot y = x_1y_1 + x_2y_2 + x_3y_3 + \ldots + x_ny_n$$

Must be conformable for multiplication One way to understand why matrices must be conformable is to try and multiply non-conformable matrices such as A and B.

$$(9)\quad \text{if } A = \begin{bmatrix} a_{11} \\ a_{21} \end{bmatrix} \text{ and } B = \begin{bmatrix} b_{11} & b_{12} \\ b_{21} & b_{22} \end{bmatrix}$$

However the product BA is conformable for multiplication. Observe how B row 1 and A column 1 form a dot product, and how B row 2 and A column 1 form a dot product

$$(10)\quad BA = \begin{bmatrix} b_{11} & b_{12} \\ b_{21} & b_{22} \end{bmatrix} \times \begin{bmatrix} a_{11} \\ a_{21} \end{bmatrix} = \begin{bmatrix} b_{11}a_{11} + b_{12}a_{21} \\ b_{21}a_{11} + b_{22}a_{21} \end{bmatrix}$$

An elementary view of matrix multiplication Dashes are don't cares.

$$(11a)\quad \begin{bmatrix}1 & 2\\ - & -\end{bmatrix}\times\begin{bmatrix}4 & -\\ 6 & -\end{bmatrix}=\begin{bmatrix}1\times4+2\times6 & -\\ - & -\end{bmatrix}=\begin{bmatrix}16 & -\\ - & -\end{bmatrix}$$

$$(11b)\quad \begin{bmatrix}- & -\\ 3 & 4\end{bmatrix}\times\begin{bmatrix}4 & -\\ 6 & -\end{bmatrix}=\begin{bmatrix}- & -\\ 3\times4+4\times6 & -\end{bmatrix}=\begin{bmatrix}- & -\\ 36 & -\end{bmatrix}$$

$$(11c)\quad \begin{bmatrix}1 & 2\\ - & -\end{bmatrix}\times\begin{bmatrix}- & 5\\ - & 7\end{bmatrix}=\begin{bmatrix}- & 1\times5+2\times7\\ - & -\end{bmatrix}=\begin{bmatrix}- & 19\\ - & -\end{bmatrix}$$

$$(11d)\quad \begin{bmatrix}- & -\\ 3 & 4\end{bmatrix}\times\begin{bmatrix}- & 5\\ - & 7\end{bmatrix}=\begin{bmatrix}- & -\\ - & 3\times5+4\times7\end{bmatrix}=\begin{bmatrix}- & -\\ - & 43\end{bmatrix}$$

Linear substitution is also a guide to matrix multiplication. Consider the equations where a's and b's are constants.

(12a) $x_1 = a_{11}y_1 + a_{12}y_2$ (13a) $y_1 = b_{11}z_1 + b_{12}z_2$

(12b) $x_2 = a_{21}y_1 + a_{22}y_2$ (13b) $y_2 = b_{21}z_1 + b_{22}z_2$

(12c) $X = AY$ (13c) $Y = BZ$

(14a) $x_1 = a_{11}(b_{11}z_1 + b_{12}z_2) + a_{12}(b_{21}z_1 + b_{22}z_2)$

(14b) $x_2 = a_{21}(b_{11}z_1 + b_{12}z_2) + a_{22}(b_{21}z_1 + b_{22}z_2)$

(15a) $x_1 = (a_{11}b_{11} + a_{12}b_{21})z_1 + (a_{11}b_{12} + a_{12}b_{22})z_2$

(15b) $x_2 = (a_{21}b_{11} + a_{22}b_{21})z_1 + (a_{21}b_{12} + a_{22}b_{22})z_2$

(15c) $X = ABZ$

Two matrices A and B are conformable for multiplication as AB when number of columns in A equals the number of rows in B. In turn AB and Z are conformable for multiplication.

The element a_{ij} subscripts are row and column numbers. The first number is a row number. The second number is a column number.

Two matrices A, B are equal, and we write A=B, when the matrices are conformable and when each element of A equals the corresponding element of B.

Matrices may or may not commute Matrix AB may not equal matrix BA. For example.

$$(16a)\quad AB = \begin{bmatrix} 1 & 2 \\ 1 & 2 \end{bmatrix} \times \begin{bmatrix} 2 & 1 \\ 2 & 1 \end{bmatrix} = \begin{bmatrix} 6 & 3 \\ 6 & 3 \end{bmatrix} \qquad (16b)\quad BA = \begin{bmatrix} 2 & 1 \\ 2 & 1 \end{bmatrix} \times \begin{bmatrix} 1 & 2 \\ 1 & 2 \end{bmatrix} = \begin{bmatrix} 3 & 6 \\ 3 & 6 \end{bmatrix}$$

The unit matrix commutes The square matrix of order n that has ones in its leading diagonal and zeros elsewhere is referred to as the *unit matrix* of order n. The unit matrix symbol is I.

$$(17)\quad I = I_{5\times 5} = \begin{bmatrix} 1 & 0 & 0 & 0 & 0 \\ 0 & 1 & 0 & 0 & 0 \\ 0 & 0 & 1 & 0 & 0 \\ 0 & 0 & 0 & 1 & 0 \\ 0 & 0 & 0 & 0 & 1 \end{bmatrix}$$

$$(18a)\quad IA = AI = A \qquad (18b)\quad I = I^2 = I^3 = \ldots.$$

Distributive and associative laws for multiplication apply to matrices

$$(19a)\quad (A+B)C = AC + BC \qquad (19b)\quad (AB)C = A(BC)$$

The Division Law The division law in ordinary algebra states that when the product xy=0, either x or y is zero, or both must be zero. This law *does not apply* to matrix products. The product AB may equal zero, however this does not imply A or B are the null matrix (Problem 1202).

Cancellation If ab=ac, a≠0, in the algebra of numbers, then we may cancel a on both sides of = so that b=c. However this may not be possible in a matrix equation such as AB=AC where B≠C (Problem 1203)

Problem 1201 Let A be a 2×2 matrix. Show that 0×A=A×0=0.

Problem 1202 $A = \begin{bmatrix} a & b \\ 0 & 0 \end{bmatrix}$ $B = \begin{bmatrix} b & 2b \\ -a & -2a \end{bmatrix}$ Show that AB=0, BA≠0.

Problem 1203 $\mathrm{A} = \begin{bmatrix} 0 & 0 \\ 0 & 1 \end{bmatrix}$ $\mathrm{B} = \begin{bmatrix} 1 & 1 \\ 1 & 1 \end{bmatrix}$ $\mathrm{C} = \begin{bmatrix} 0 & 0 \\ 1 & 1 \end{bmatrix}$

Show that AB=C, AC=C , BC=B

12.3 Related Matrices

The transpose of a matrix is marked as A^T. Rows become columns.

$$(20)\quad A=\begin{bmatrix} a_{11} & a_{12} & a_{13} \\ a_{21} & a_{22} & a_{23} \end{bmatrix} \qquad A^T=\begin{bmatrix} a_{11} & a_{21} \\ a_{12} & a_{22} \\ a_{13} & a_{23} \end{bmatrix}$$

$$(21a)\quad (AB)^T = B^T A^T \qquad (21b)\quad (ABC)^T = C^T B^T A^T$$

The reciprocal or inverse of a matrix is marked as A^{-1} How to calculate the inverse is demonstrated in Section 12.5.

$$(22a)\quad AA^{-1} = I \qquad (22b)\quad A^{-1}A = I$$

$$(23a)\quad (A^{-1})^{-1} = A \quad (23b)\quad (A^{-1}B^{-1})^{-1} = BA$$

$$(24a)\quad (AB)^{-1} = B^{-1}A^{-1} \qquad (24b)\quad (ABC)^{-1} = C^{-1}B^{-1}A^{-1}$$

Reciprocating and Transposing commute

$$(25)\quad (A^T)^{-1} = (A^{-1})^T$$

Positive and Negative Integer Matrix Exponents A^r

$$(26)\quad A^2 = AA \quad A^3 = AAA \quad A^{-2} = A^{-1}A^{-1} \quad \textit{and so forth}$$

$$(27)\quad \mathrm{A^R A^S = A^{R+S}} \quad \mathrm{(A^{-1})^S = A^{-S}} \quad \mathrm{(A^R)^S = A^{RS}}$$

Non-singular Matrices B and C have an inverses B^{-1} and C^{-1}.

$$(28a)\quad \text{If } \mathrm{A = BX} \text{ then } \mathrm{B^{-1}A = B^{-1}BX} \rightarrow \mathrm{X = B^{-1}A}$$

$$(28b)\quad \text{If } \mathrm{D = YC} \text{ then } \mathrm{DC^{-1} = YCC^{-1}} \rightarrow \mathrm{Y = DC^{-1}}$$

Problem 1204 Multiply M×G to calculate the elements of matrix C.

$$C = M \times G$$

$$C = \begin{bmatrix} c_6 & c_5 & c_4 & c_3 & c_2 & c_1 & c_0 \end{bmatrix}$$

$$C = \begin{bmatrix} m_3 & m_2 & m_1 & m_0 \end{bmatrix} \times \begin{bmatrix} I_{4\times4} \mid R_{4\times3} \end{bmatrix}$$

$$C = \begin{bmatrix} m_3 & m_2 & m_1 & m_0 \end{bmatrix} \times \begin{bmatrix} 1 & 0 & 0 & 0 & 1 & 0 & 1 \\ 0 & 1 & 0 & 0 & 1 & 1 & 1 \\ 0 & 0 & 1 & 0 & 1 & 1 & 0 \\ 0 & 0 & 0 & 1 & 0 & 1 & 1 \end{bmatrix}$$

12.4 Rank of a Matrix

Definition of Rank A matrix has rank R when R is the largest integer for which *not all minors of order R are zero.*

Minor of order R The elements of a matrix are minors of order 1. They are determinants of order 1.

A minor of order R has a determinant of order R.

Linear Dependence If r_p and r_q represent two rows of a matrix that has c columns, then $r_p + r_q$ is the c sums, element by element, of the two rows.

(29) If n_p, n_q, n_t are any three numbers so that $n_p r_p + n_q r_q + n_t r_t = 0$, then the three rows r_p r_q r_t are linearly dependent.

Rank The rank of a matrix is equal to the number of linearly *independent* rows in the matrix.

(30) *If matrix A has n rows and rank $r < n$ then there are $n - r$ linearly dependent rows that can be expressed as sums of the r linearly independent rows.*

The rank of a matrix is difficult to determine. Determining the rank is difficult in the sense the calculations can be extensive, and difficult in the sense of finding a general format.

On the other hand there are special matrices such as a Vandermonde matrix, which has a special form *whose rank is known*. The elements of the rows are powers of the elements y_k in the first row.

$$(31)\quad V_{n \times n} = \begin{bmatrix} y_1 & y_2 & \cdots & y_n \\ y_1^{\,2} & y_2^{\,2} & \cdots & y_n^{\,2} \\ \vdots & \vdots & & \vdots \\ y_1^{\,n} & y_2^{\,n} & \cdots & y_n^{\,n} \end{bmatrix} \quad rank = n, \text{ all } y_i \text{ are distinct \& non zero}$$

Vandermonde matrices solve an important problem in Error Correction Code design.

12.5 Determinants of a Matrix

The determinant of a matrix A is

$$(32)\quad A=\begin{bmatrix} a_{11} & a_{12} & a_{13} \\ a_{21} & a_{22} & a_{23} \\ a_{31} & a_{32} & a_{33} \end{bmatrix} \rightarrow \Delta=\begin{vmatrix} a_{11} & a_{12} & a_{13} \\ a_{21} & a_{22} & a_{23} \\ a_{31} & a_{32} & a_{33} \end{vmatrix} \quad \textit{the determinant of } A$$

Minor Δ_{pq} is formed by striking out row p and column q. The expansion by minors creates the n! terms of a n×n determinant. There are 6 terms (3!) in a 3×3 determinant. For example:

$$(33)\quad \Delta = a_{11}\Delta_{11} - a_{12}\Delta_{12} + a_{13}\Delta_{13} \quad (\textit{expansion by row } 1)$$
$$\Delta = a_{11}(a_{22}a_{33} - a_{23}a_{32}) - a_{12}(a_{21}a_{33} - a_{23}a_{31}) + a_{13}(a_{21}a_{32} - a_{22}a_{31})$$

The inverse of A is as follows so that $AA^{-1}=A^{-1}A=I$. Observe that row expansions are *columns* in A^{-1}.

$$(34)\quad A^{-1}=\begin{bmatrix} \frac{\Delta_{11}}{\Delta} & \frac{-\Delta_{21}}{\Delta} & \frac{\Delta_{31}}{\Delta} \\ \frac{-\Delta_{12}}{\Delta} & \frac{\Delta_{22}}{\Delta} & \frac{-\Delta_{32}}{\Delta} \\ \frac{\Delta_{13}}{\Delta} & \frac{-\Delta_{23}}{\Delta} & \frac{\Delta_{33}}{\Delta} \end{bmatrix}$$

The product AA^{-1} confirms the A^{-1} format.

$$(35)\quad AA^{-1}=\begin{bmatrix} a_{11} & a_{12} & a_{13} \\ a_{21} & a_{22} & a_{23} \\ a_{31} & a_{32} & a_{33} \end{bmatrix} \times \begin{bmatrix} \frac{\Delta_{11}}{\Delta} & \frac{-\Delta_{21}}{\Delta} & \frac{\Delta_{31}}{\Delta} \\ \frac{-\Delta_{12}}{\Delta} & \frac{\Delta_{22}}{\Delta} & \frac{-\Delta_{32}}{\Delta} \\ \frac{\Delta_{13}}{\Delta} & \frac{-\Delta_{23}}{\Delta} & \frac{\Delta_{33}}{\Delta} \end{bmatrix}$$

$$=\begin{bmatrix} a_{11}\frac{\Delta_{11}}{\Delta}-a_{12}\frac{\Delta_{12}}{\Delta}+a_{13}\frac{\Delta_{13}}{\Delta} & 0 & 0 \\ 0 & -a_{21}\frac{\Delta_{21}}{\Delta}+a_{22}\frac{\Delta_{22}}{\Delta}-a_{23}\frac{\Delta_{23}}{\Delta} & 0 \\ 0 & 0 & a_{31}\frac{\Delta_{31}}{\Delta}-a_{32}\frac{\Delta_{32}}{\Delta}+a_{33}\frac{\Delta_{33}}{\Delta} \end{bmatrix}=\begin{bmatrix} 1 & 0 & 0 \\ 0 & 1 & 0 \\ 0 & 0 & 1 \end{bmatrix}=I$$

Trigonometry

1 Angles and their Measure

Angles of all magnitudes, positive and negative, appear in scientific and engineering problems. Trigonometry includes the study and measure of angles. Our immediate concern is with *plane angles*.

Plane Angles A plane angle is generated by rotating a *generating line* in a plane about a point O referred to as a *vertex* (Figure 101). The initial position OA of the generating line is called the *initial side* of the angle, and its final position OB is called the *terminal side*. If the rotation is counterclockwise the angle is referred to as positive, if clockwise, it is referred to as negative.

The generation of an angle Let the initial side be a fixed straight line on the x axis, and let the terminal side be a straight line coincident with the initial side (Figure 101). Rotate the terminal side around the vertex in the counterclockwise direction to generate the included angle

Figure 101

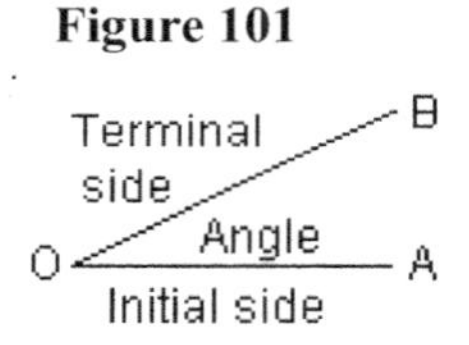

A common way to measure angles is the system of degrees °. A full revolution about a point is divided into 360 parts of equal size where each part is designated as one degree. An angle measuring 180° is a straight line. A 90° angle forms a *right angle* such as the angles found at the corners of a square. *Acute angles* have measure less than 90° and *obtuse angles* are angles that have measure between 90° and 180° (Figure 102).

Figure 102

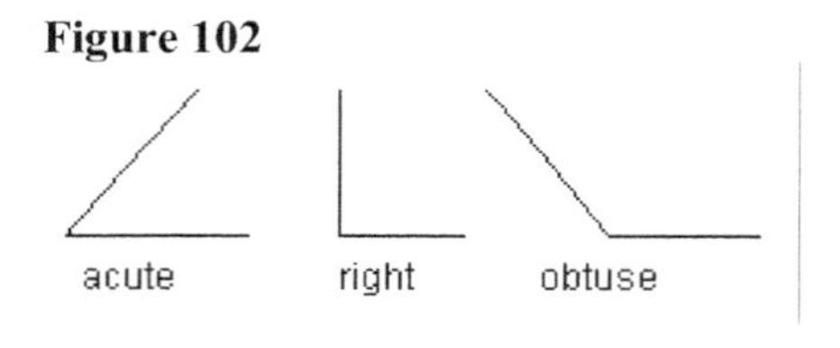

Important Angles may be associated in pairs (Figure 103). *Complementary* angles are angles whose measures add to 90°. *Supplementary* angles are angles whose measures add to 180°.

Figure 103

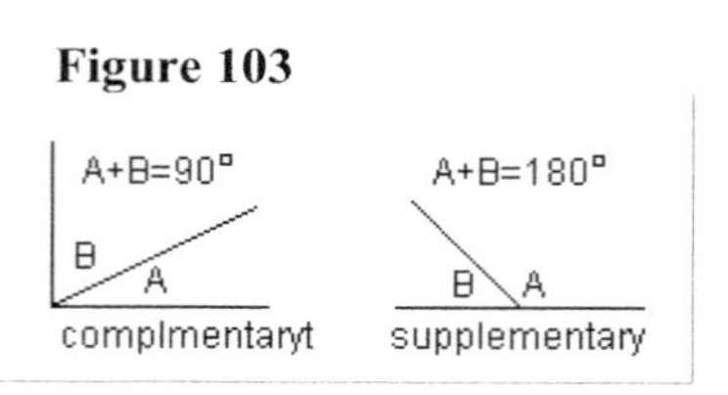

Two angles that have the same measure are referred to as *congruent* angles.

1.1 The Degree

The next step is to decide upon a system for the numerical measurement of angles. In order to do this, the world decided on a unit angle, which is an arbitrarily chosen angle of fixed magnitude. Then all other angles are measured numerically by their ratios to this unit angle. In practice there are two units for angles. One unit is the degree, which is defined as 1/360 of one revolution around a point. The other unit is the radian where 2π radians around a point represents one revolution.

A *degree* is a unit of measure for angles. One degree is the measure of an angle with its vertex at the center of a circle, and with its sides marking 1/360 of the circumference of the circle.

The degree is subdivided into sixty parts called minutes, the minute is subdivided into sixty parts called seconds. This is the sexagesimal system of numerical measurement of angles. An angle is written as d°m's" degrees, minutes, seconds.

$$angle\ d^\circ m's'' = d + \frac{m}{60} + \frac{s}{60^2}\ degrees$$

Angles smaller than a second are written as decimals of a second.

We prefer the all decimal representation.

Problem 101 Draw figures showing rotation producing angles of 90°, 315°, –270°, –45°, 1100°.

1.2 The Radian

A *radian* is the other unit of measure for angles. One radian is the measure of an angle with its vertex at the center of a circle, and with sides cutting off an arc equal in length to the radius of the circle (Figure 104). We learned in geometry that the arc length of a complete circle is $2\pi r$, and so 2π radians is the angle marked by a complete circle. (For theoretical purposes the radian turns out to be more convenient to use.)

Figure 104

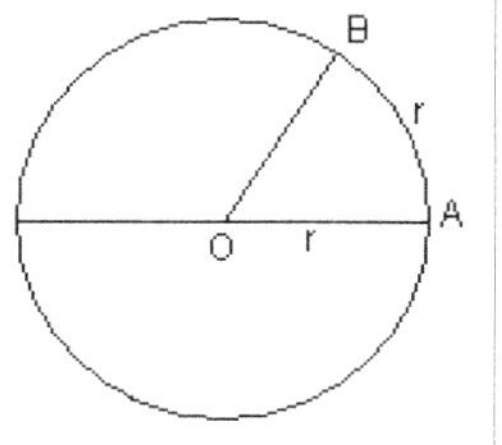

Euclid showed us the length of the circumference of a circle is $2\pi r$. The ratio of the circumference to its radius r is the irrational number 2π.

$$\pi = \frac{\text{circumference of a circle}}{2 \times \text{radius of a circle}} = \frac{\text{circumference of a circle}}{\text{diameter of a circle}}$$

Once around a circle spans an angle of 2π radians. A half circle is π radians and so π radians equal 180°.

A number is irrational when the ratio m/n is not an integer. The number π equals an infinite non-recurring decimal 3.14159265358979323846 2....

1.3 Arc Length

Arc AB is the distance *on the circumference of a circle* from point A to point B (Figure 105). Consequently arc AB is proportional to angle AOB.

$$(1) \quad \frac{arc\ AB}{angle\ AOB} = \frac{circumference\ of\ the\ circle}{2\pi\ radians} \quad \Rightarrow \quad \frac{a}{\theta} = \frac{2\pi r}{2\pi} \quad \Rightarrow \quad a = r\theta$$

Observe that the magnitude of angle AOB is independent of the radius r (Figure 105). If the angle AOB measure is one radian, then on the three circles the ratio of the arc length r to the radius r is the same, i.e. 1.

Figure 105

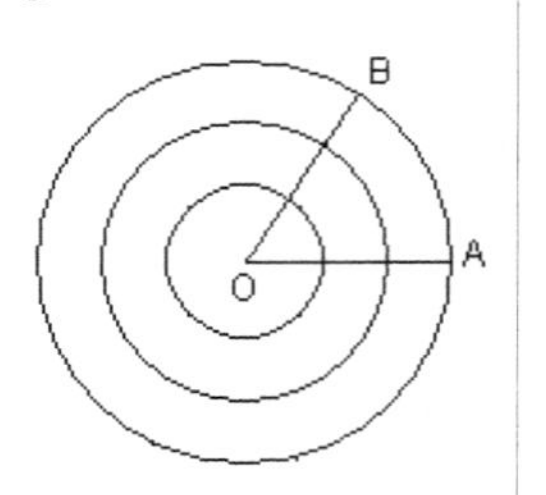

1.4 Relate Degrees to Radians

$$(2) \quad 2\pi\,\text{radians} = 360° \quad \rightarrow \quad 1\,\text{radian} = 360° \frac{1}{2\pi} = 57.2958°$$

1.5 Area of a Sector of a Circle

Euclidean geometry showed us that the area of a circle is πr^2. The area of the sector AOB of a circle is proportional to the included angle θ.

$$(3) \quad \frac{area\ of\ AOB}{\pi r^2} = \frac{\theta}{2\pi} \quad \Rightarrow \quad area\ of\ AOB = \frac{r^2\theta}{2}$$

Problem 102 Show that the radian measure of a 60° angle is $\pi/3$.

Problem 103 Show that the length of an arc defined by a 2.58 radian angle is 2.58 times the radius r.

Problem 104 Show that the radian measure of 1/4 of a circle is $\pi/2$.

Problem 105 Show that the radian measure of a 45° angle is $\pi/4$.

Problem 106 Show that the degree measure of a π radian angle is 180°.

Problem 107 Show that the degree measure of a $\pi/6$ radian angle is 30°.

Problem 108 Show that the radian measure of a 1° angle is $\pi/180$.

Problem 109 Show that the axis of a wheel of radius 1 meter will move forward 2π meters for each revolution.

Problem 110 Show that when the axis of a wheel of radius 25cm moves forward 2 meters it rotates through 8 radians.

Problem 111 Given the length of the arc a and the radius r show that, the radian angle subtended at the center is (a) 0.128 and (b)1.054 where (a) $a = 0.16$, $r = 1.25$ (b) $a = 1.36$, $r = 1.29$.

Problem 112 On a circle of radius 5.782 meters the length of an arc is 1.742 meters. Show that angle subtended at the center is 0.301 radians

Problem 113 Show that the change in latitude by walking due north one mile, assuming the earth to be a sphere of radius 3956 miles is an angle of 0.0002528 radians.

Problem 114 Show that the length in feet of one minute of arc on a great circle of the earth, radius 3956 miles, is 0.0002909 radians or 967 feet.

Significant Digits

When are Digits Significant? Non-zero digits are always significant. The number 54 has two significant digits, and 54.7 has three significant digits. An integer has an infinite number of significant digits (e.g. 33.0000....).

Digits that are zeros involve several "rules"
1. Zeros before other digits are not significant. In 0.046 the first two zeros are not significant. The number has two significant digits.
2. Zeros between non zero digits are always significant. The number 3008 has four significant digits.
3. Zeros after other digits and following a decimal point are significant. The number 2.650 has four significant digits.
4. Zeros at the end of a number are significant only if they follow a decimal point as in item 3. Otherwise, it is impossible to tell if they are significant. In the number 250,000 we cannot say how many zeroes are significant. The 2 and 5 digits are significant digits. If all the zeros of 250,000 are significant, then use scientific notation where significant zeroes follow a decimal point. For example, numbers such as these have 6, 5, and 2 significant digits.

2.50000×10^{5} (6) $\quad$ 2.5000×10^{5} (5) $\quad$ 2.5×10^{5} (2)

Operations The number of significant digits in an answer produced by *multiplication or division* equals the smallest number of significant digits in any one of the numbers involved. For example, if x=0.351 (3 significant digits) and y=5.241 (4 significant digits) the product xy and trigonometric function tan(xy) have 3 significant digits.

The number of significant digits in an answer produced by *addition or subtraction* depends on the number of digits before and after the decimal points. The number of significant digits after the decimal points is the smallest number of such digits in the numbers (the 0.1). The number of significant digits before the decimal points is the greatest number of such digits in any one number (the 1342).

$$\begin{array}{r} 1342.021 \\ 1.0460 \\ +\ 32.1 \\ \hline 1375.1 \end{array}$$

Retain at least one extra digit in intermediate answers than will be needed in a final answer. If all intermediate answers are rounded-off to fewer digits, then information is lost. The last digit(s) in a final answer may be wrong. Avoid these "round-off errors."

2 Trigonometric Functions

One way to define trigonometric functions is as ratios of two sides of a right triangle. There are three sides in a triangle, which means there are six ways to form ratios. That is why there are six trigonometric functions. The sine and cosine functions are the basic functions. The other four functions are derived from the sine and/or cosine.

Figure 201 The right triangle

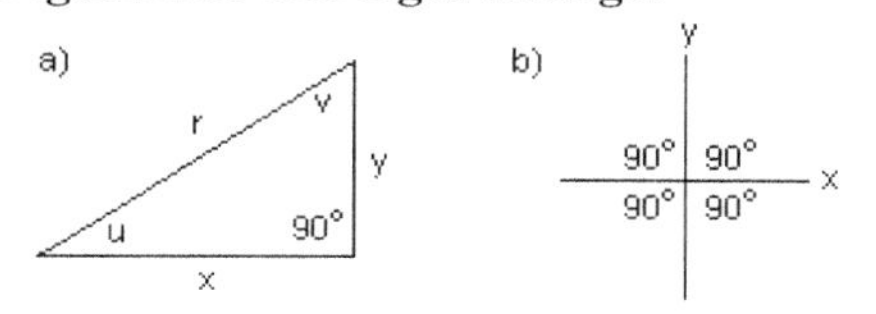

The triangle with sides r, y, x is a *right triangle* when x and y are perpendicular (Figure 201a). The angle defined by sides x and y is 90° (Figures 201a, 201b).

We start by *defining* sin u and cos u.

Sine

The sine of angle u is *defined* as the ratio y/r.

(1) $\sin u = \dfrac{y}{r} = \dfrac{opposite\ side}{hypotenuse}$

Cosine

The cosine of angle u is *defined* as the ratio x/r.

(2) $\cos u = \dfrac{x}{r} = \dfrac{adjacent\ side}{hypotenuse}$

Angles u and v are complementary (Figure 201)

(3*a*) $u + v = \dfrac{\pi}{2} \rightarrow \sin u = \dfrac{y}{r} = \cos v \rightarrow \sin u = \cos(\pi/2 - u)$

(3*b*) $\cos u = \dfrac{x}{r} = \sin v \rightarrow \cos u = \sin(\pi/2 - u)$

2.1 Sine and Cosine Waveforms

Rotation on the unit circle Moving once around the circle, a particle P rotates through an angle of 2π *radians*. Regarding the particle, we ask what is the variation in the particle's x and y coordinates as the particle rotates around the circle? How do x and y vary as a function of particle position angle u whose dimension is in radians?

Figure 202

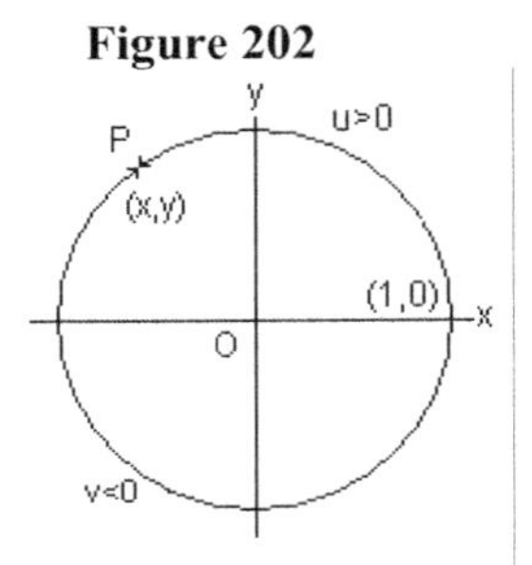

Changes in y as the particle moves around the circle At (1, 0) y equals 0. As angle u increases y increases, reaching a maximum value of 1 at u=π/2 (Figure 202). Then y decreases again to 0 as u approaches π. Moving u past π increases y in a negative direction, reaching a maximum value of −1 at u=3π/2. Then y decreases to 0 as u approaches 2π (returns to 0).

Changes in x as the particle moves around the circle At (1, 0) x equals 1. As u increases x decreases, reaching 0 at u=π/2. Then x increases in a negative direction, reaching a maximum value of −1 at u=π. Moving x past π decreases x to 0 as u approaches 3π/2. Then x increases to 1 as u approaches 2π (returns to 0).

Figure 203 Waveforms of sin u and cos u

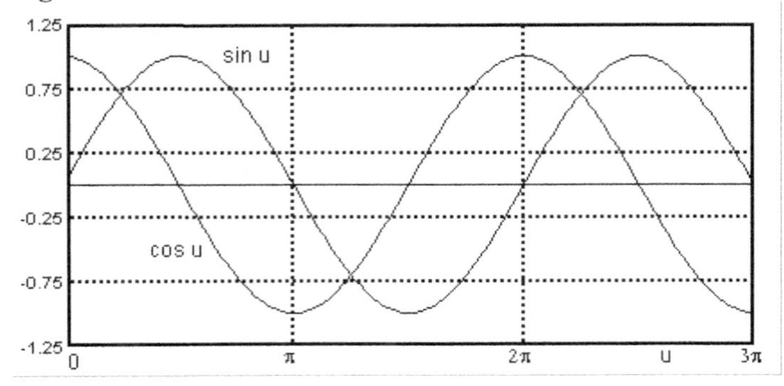

Waveforms of sin u and cos u Plot the values of coordinates x and y as the particle rotates counterclockwise around the circle (from u=0 to u=2π). Draw a smooth curve through the y points. This is the *waveform* for sin u (Figure 203). Repeat the x points to create the waveform for cos u. The range is from −1 to +1. The waveforms are identical except for a π/2 shift.

2.2 The Sine and Cosine Functions

The basic periodic functions are sin u and cos u. (The cos u properties are the same as sin u except for a $\pi/2$ shift.) Periodic because each time u increases by 2π the sin u and cos u waveforms repeat (Figure 203, 204). The 2π interval is referred to as the period of sin u and cos u. Here angle u is measured in radians.

The basic general periodic function is y=sin x where y and x represent real numbers, and the fact that x represents angles is irrelevant. Note that we can change the amplitude from 1 to 2 (Figure 204) or halve the period as well as change the amplitude from 1 to 2 (Figure 205).

Figure 204 The waveforms for $V_2 = 2\sin x$ and $V_1 = \sin x$

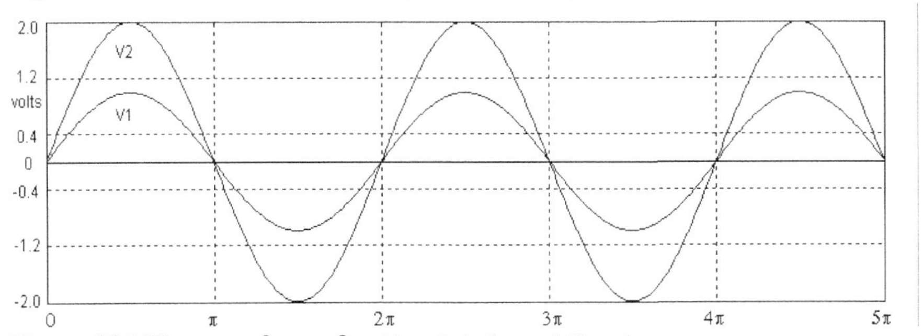

Figure 205 The waveforms for $V_2 = 2\sin 2x$ and $V_1 = \sin x$

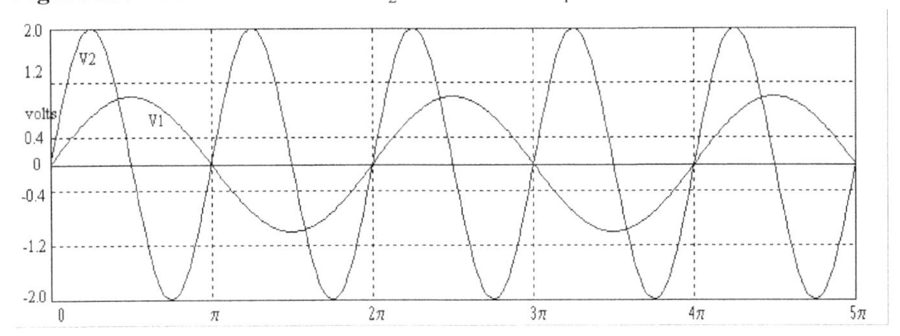

Trigonometric functions can be functions of time, distance, whatever.

$$(4a) \quad y = \sin x = \sin 2\pi \frac{t}{T} \qquad (4b) \quad y = \sin x = \sin 2\pi \frac{l}{L}$$

Emphasis Trigonometric functions are periodic. This is why they are immensely important.

2.3 The Other Four Functions

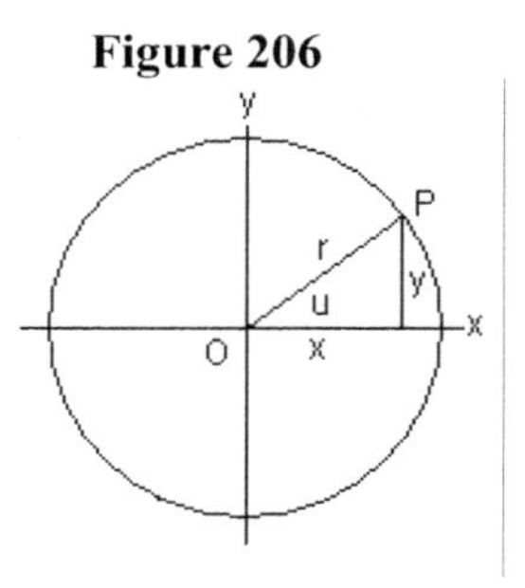

Figure 206

Sine and cosine functions are defined by the ratios y/r and x/r respectively (Figure 206). The *tangent* function does not involve r, because it is defined by the ratio y/x. Three other functions (*cosecant, secant, and cotangent*) are simply reciprocals of sine, cosine, and tangent. Their definitions are shown in Table 201.

Table 201

sine	$\sin u = \frac{y}{r}$	*cosecant*	$\csc u = \frac{r}{y} = \frac{1}{\sin u}$
cosine	$\cos u = \frac{x}{r}$	*secant*	$\sec u = \frac{r}{x} = \frac{1}{\cos u}$
tangent	$\tan u = \frac{y}{x} = \frac{\sin u}{\cos u}$	*cotangent*	$\cot u = \frac{x}{y} = \frac{1}{\tan u}$

2.3.1 The Tangent

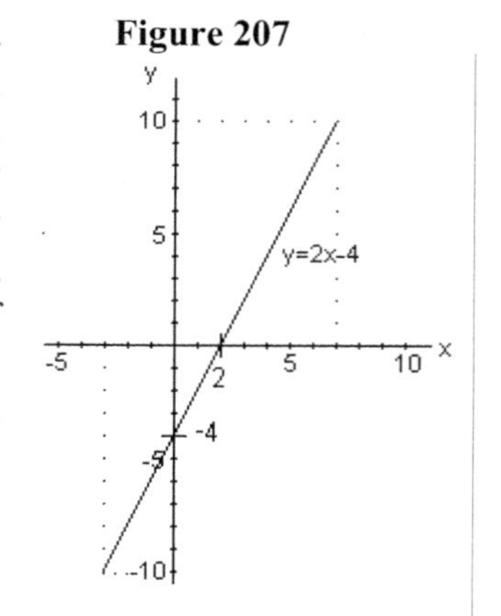

Figure 207

A simple example of the tangent is found in a straight line. The equation of a straight line in the x, y plane is y=mx+b. Select a value of x, calculate the value of y, and plot the point x, y (Figure 207). Observe that incrementing x by 1 increments y by 2 (equation 5). This is a consequence of the term mx=2x. The ratio of increments is the *slope* of the line, which is equal to m. The increment in x is multiplied by m, which produces the increment in y. The slope m is also referred to as the *tangent*. Note that when x = 0, y = b = –4, and when y = 0, x= –b/m = 2.

(5) $y = mx + b = 2x - 4$

x	−3	−2	−1	0	1	2	3	4	5	6	7	....
y	−10	−8	−6	−4	−2	0	2	4	6	8	10	....

Tangent in terms of sine and cosine The tangent of angle u is defined as the increment in y produced by the increment in x (Figure 206).

Figure 206

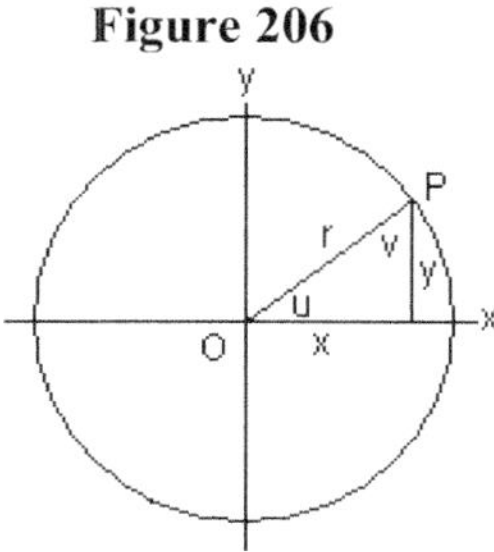

$$(6) \quad \tan u = m = \frac{\Delta y}{\Delta x} = \frac{y}{x} = \frac{y/r}{x/r} = \frac{\sin u}{\cos u}$$

Observe that as angle u approaches π/2, x goes to zero, and the ratio y/x=tan u increases towards infinity (a number that is as large as we please).

2.3.2 Relations between the Functions

The intent here is to show how the Pythagorean Identity (Chapter 3) is used to create relationships among the trigonometric functions. There is no need to memorize formulas when an identity can be used to create a formula when needed.

Dotted lines are added to Table 202 to emphasize that when the equation at the head of a row or column is true, the row or column entries are true. For example when radius r=1

$$(7a) \quad \textit{If} \ \tan u = t, \ \sin u = s, \ \cos u = c, \ \textit{then} \ t = \frac{s}{c} \ \textit{and} \ s^2 + c^2 = 1$$

$$(7b) \quad s^2 = c^2 t^2 = (1 - s^2) t^2 \quad \rightarrow \quad s^2 (1 + t^2) = t^2 \quad \rightarrow \quad s = \frac{t}{\sqrt{1 + t^2}} = \frac{x}{r}$$

Table 202

*Figure*206	$\sin u$	$\cos u$	$\tan u$
$\sin u$	$\sin u$	$\sqrt{1-\sin^2 u}$	$\frac{\sin u}{\sqrt{1-\sin^2 u}}$
$\cos u$	$\sqrt{1-\cos^2 u}$	$\cos u$	$\frac{\sqrt{1-\cos^2 u}}{\cos u}$
$\tan u$	$\frac{\tan u}{\sqrt{1+\tan^2 u}}$	$\frac{1}{\sqrt{1+\tan^2 u}}$	$\tan u$

Rectangular coordinates In order to describe the position of a point in a plane, two reference lines, the x axis and the y axis, are chosen at right angles to each other. Their point of intersection O is called the origin (Figure 206). The position of any point in the plane, such as P, is described by specifying its distance from the origin on the x axis, and its distance from the origin on the y axis. These distances are called the x and y coordinates of the point. The coordinates of any point, such as P, are written as (x, y), and the coordinates of the "start point" at the intersection of the unit circle and the positive x axis are (1, 0). Note that the *signs* of x and y may be plus or minus as point P rotates around the circle.

Complementary angle The sine of angle u equals the cosine of the complementary angle v (Figure 206). The complementary angle v equals 90° minus angle u. For example, if u = 33°, then the complement v = 57°.

$$(8a)\quad \cos v = \frac{y}{r} = \sin u \;\; \textit{and} \;\; u + v = 90° = \frac{\pi}{2}$$

$$(8b)\quad \cos v = \sin\left(\frac{\pi}{2} - v\right)$$

Figure 206

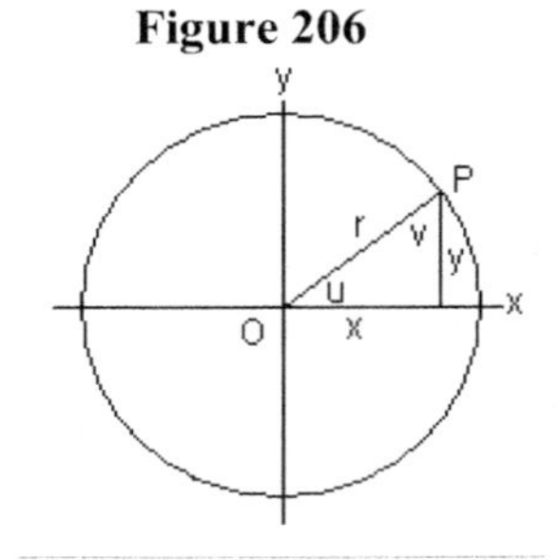

sin u and cos u are periodic A function f(z) is referred to as *periodic*, when f(z) of variable z has the property f(z)=f(z+k) for every value of z and k is a constant. Furthermore k is referred to as the *period* of the function, when k is the smallest constant for which f(z) has this property. If f(z)=f(z+k), then f(z)=f(z+nk) where n is any positive or negative integer. I.e.

$$(9a)\quad \textit{If } f(z) = f(z+k), \;\; \textit{then } f(z+k) = f(z+2k), \cdots, \textit{so that}$$

$$(9b)\quad f(z) = f(z+k) = f(z+2k) = \cdots = f(z+nk)$$

Clearly the sine and cosine functions repeat each time the circle is traversed. In other words sin and cos are periodic, the period is 2π, and so k=2π. I.e.

$$(10a)\quad f(x) = \sin x = \sin(x+2\pi) = \cdots = \sin(x+2n\pi)$$

$$(10b)\quad g(x) = \cos x = \cos(x+2\pi) = \cdots = \cos(x+2n\pi)$$

Confusion is avoided if you focus on the *sign* of x and y.

Use a calculator if necessary.
Problem 201 In a right triangle the hypotenuse r=15 cm, and sin u=2/5 (Figure 201). Find y and x.

Problem 202 In a right triangle v=55.5°, and x= 6. Find u and y.

Problem 203 In a right triangle the hypotenuse r = 1 meter, and sin u = 0.6. Show that x = 0.8 meter and y = 0.6 meter.

Problem 204 In a right triangle cos u =0.180, r=1.000. Show that x=0.18, y=0.984, and u=79.63°. Verify that $x^2+y^2=1$

Problem 205 In a right triangle u = 55.5°, and y = 6.05. Show that the hypotenuse r = 7.34 and x = 4.16.

Problem 206 In a right triangle the angles are u, π/2–u, and π/2. Let r=1. Make sketches that show that sin u=y, cos u=x, sin(π/2–u)=x, cos(π/2–u)=y, sin(π/2)=1, cos(π/2)=0.

Problem 207 In this triangle a and c form a π/2 angle, and so do b and d. Find three ratios each for sin u, cos u, sin v, cos v.

Figure p208

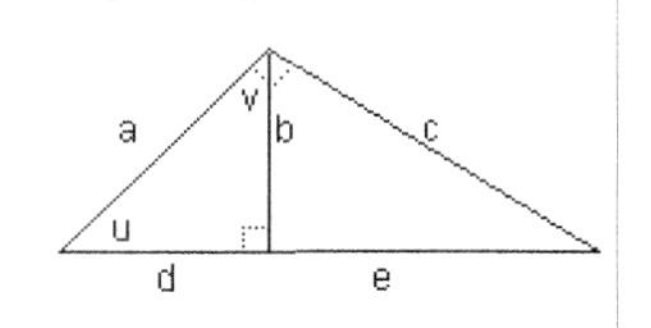

Problem 208 Plot the line whose equation has slope m=1/2, and intercept b=5/2

Problem 209 If sin u=x find cos u and tan u.

Problem 210 If cos u=x find sin u and tan u.

Problem 211 If tan u=x find cos u and sin u.
Problem 212 Show that tan u = cot(90°–u) = cot(π/2–u).
Problem 213 Show that cot u = tan(90°–u) = tan(π/2–u).
Problem 214 Show that sec u = csc(90°–u) = csc(π/2–u).
Problem 215 Show that csc u = sec(90°–u) = sec(π/2–u).
Problem 216 Show that tan(-u)=–tan u, cot(-u)= –cot u,

3 Pythagorean Identities

The Pythagorean theorem: $C^2=A^2+B^2$ where C is the hypotenuse of a right triangle with sides A and B.

One Proof of the Pythagorean Theorem Draw right triangle ABC. Use four copies of right triangle ABC to form a square with sides A+B in length (Figure 301). Calculate area of the square two ways.

$$Area = (A+B)^2 = C^2 + 4\times\frac{AB}{2}$$

$$A^2 + 2AB + B^2 = C^2 + 2AB$$

$$(1)\quad A^2 + B^2 = C^2$$

Figure 301

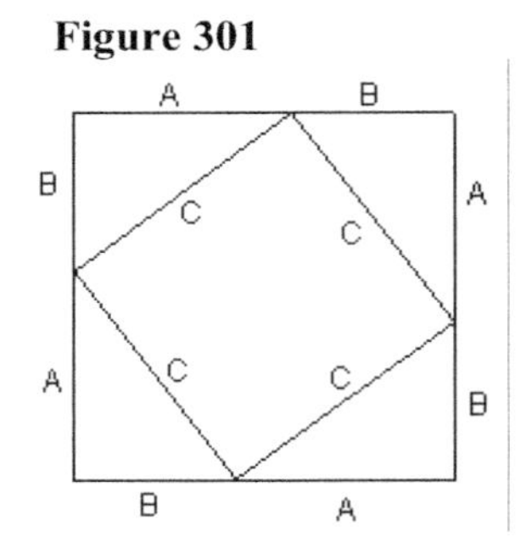

A right triangle has hypotenuse r, sides x, y and included angle u. Divide the Pythagorean equation by r^2 (2b). Substitute sine u and cos u for x and y (2c). This action produces the trigonometric form of the Pythagorean Identity (2d).

$$(2a)\quad x^2 + y^2 = r^2$$

$$(2b)\quad \frac{x^2}{r^2} + \frac{y^2}{r^2} = \frac{r^2}{r^2} = 1$$

$$(2c)\quad (\sin u)^2 + (\cos u)^2 = 1$$

$$(2d)\quad \sin^2 u + \cos^2 u = 1$$

Divide equation 2d by cosine squared and then sine squared to get the tangent and cotangent identities.

$$(3a)\quad \frac{\sin^2 u}{\cos^2 u} + \frac{\cos^2 u}{\cos^2 u} = \frac{1}{\cos^2 u}$$

$$(3b)\quad \tan^2 u + 1 = \sec^2 u$$

$$(4a)\quad \frac{\sin^2 u}{\sin^2 u} + \frac{\cos^2 u}{\sin^2 u} = \frac{1}{\sin^2 u}$$

$$(4b)\quad 1 + \cot^2 u = \csc^2 u$$

Three numbers defining a right triangle is defined as a Pythagorean triple. For any m, n ($m^2 - n^2$, *2mn*, $m^2 + n^2$) is a Pythagorean triple forming the three sides of a right triangle. For example

n	*m*	m^2-n^2	*2mn*	m^2+n^2	*check*
1	2	3	4	5	9+16=25
1	√3	2	2√3	4	4+12=16
√2	√3	1	2√6	5	1+24=25
		1	1	√2	1+1=2
		1	√3	2	1+3=4

4 Right Triangles

4.1 Solutions

Right triangle solutions are straightforward when the Pythagorean Identity (Chapter 3 page 94) and inverse trigonometric functions (Chapter 7 page 106) are used.

Figure 401

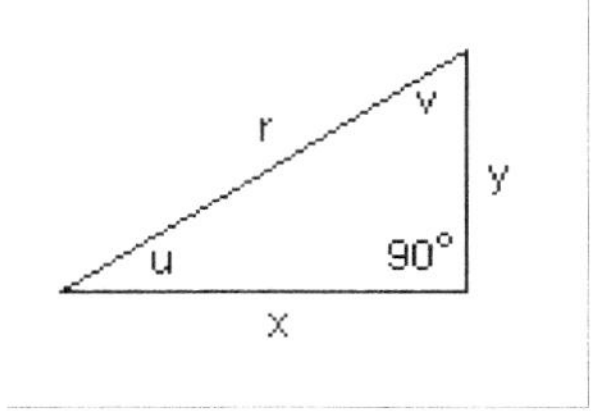

Two Sides x, y

$$(1a)\quad given\ x\ and\ y\ find\ r\ \rightarrow\ r^2 = x^2 + y^2$$

$$(1b)\quad angle\ u = \arcsin\frac{y}{r} \qquad angle\ v = \arcsin\frac{x}{r}$$

One Side x and the Hypotenuse r

$$(2a)\quad given\ x\ and\ r\ find\ y\ \rightarrow\ y^2 = r^2 - x^2$$

$$(2b)\quad angle\ u = \arcsin\frac{y}{r} \qquad angle\ v = \arcsin\frac{x}{r}$$

One side x and One Angle u

$$(3a)\quad given\ x\ and\ u\ find\ r\ \rightarrow\ r = \frac{x}{\cos u}$$

$$(3b)\quad y = r\sin u \qquad angle\ v = \frac{\pi}{2} - u$$

Hypotenuse r and One Angle u

$$(4a)\quad given\ r\ and\ u\ find\ x\ \rightarrow\ x = r\cos u$$

$$(4b)\quad y = r\sin u \qquad angle\ v = \frac{\pi}{2} - u$$

4.2 Values of sin u and cos u for special angles

The right angle isosceles triangle has angles of π/4, π/4, π/2 and hypotenuse √2 (Figure 402). The equilateral triangle has three angles equal to π/3. A perpendicular of length √3 divides the equilateral triangle into two right angle triangles with angles π/6, π/3, and π/2.

Figure 402 Special Triangles

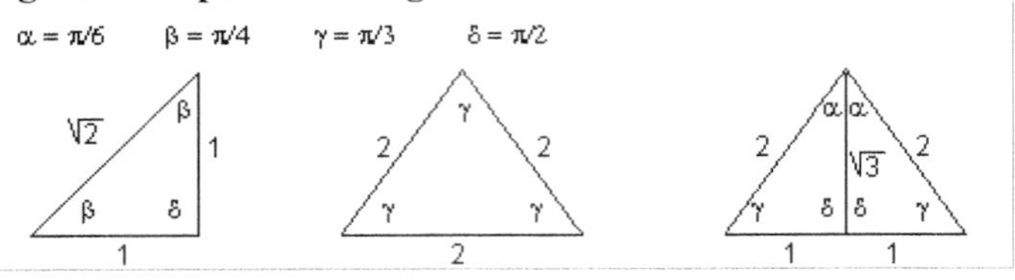

$$(5a)\ \sin\beta = \sin\frac{\pi}{4} = \frac{1}{\sqrt{2}} = \cos\frac{\pi}{4} \qquad (5b)\ \sin\gamma = \sin\frac{\pi}{3} = \frac{\sqrt{3}}{2} = \cos\frac{\pi}{6}$$

$$(5c)\ \sin\alpha = \sin\frac{\pi}{6} = \frac{1}{2} = \cos\frac{\pi}{3}$$

When γ decreases to 0 or increases to π/2 limit values are reached.

$$(6a)\quad \sin 0 = 0 = \cos\frac{\pi}{2} \qquad (6b)\ \sin\frac{\pi}{2} = 1 = \cos 0$$

Emphasis: Complementary angles Observe that sin(u)=cos(π/2–u) in the equations above. If angles α+γ= π/2=90°, then they are referred to as complementary angles (Figure 402).

Problem 401 In the following right triangles two sides are given (Figure 401). Compute the sin, cos, and tan of angles u and v. Show the results as fractions. (a) $r = 5$, x = 4 (b) r = 2, $y = \sqrt{3}$ (c). $r = 5, x = 1$.

Problem 402 In a right triangle $r = 6.25$ ft and tan $u = 1.2$. Show that x and y are 4 ft and 4.8 ft.

Problem 403 $a = 4$, $b = 4.8$. Show that r=6.25 and u=39.8°.

---

5 Triangle Laws

The three types of triangles are shown in Figure 501. They are
1) acute - three acute angles each < 90° ($\pi/2$)
2) right angle - one angle is 90° ($\pi/2$)
3) obtuse - one angle is > 90° ($\pi/2$)

Figure 501 Three types of Triangles

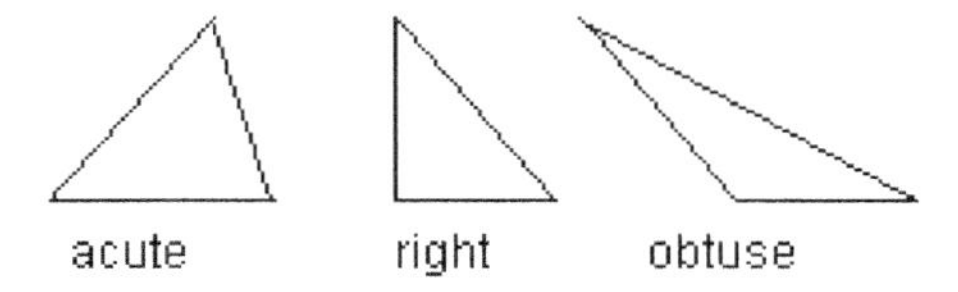

An oblique triangle is a triangle that is *not* a right triangle. An *oblique* triangle is an *acute* triangle, or an *obtuse* triangle. The laws of sines and cosines are used to solve problems involving oblique triangles.

5.1 The Law of Sines

We derive the Law of Sines using obtuse triangles, because they are complex compared to the other types. Conventional labels of parts of obtuse triangles are as follows (Figure 502). The angles are A, B, C , and the sides opposite the angles are a, b, c respectively.

Figure 502 Obtuse Triangles

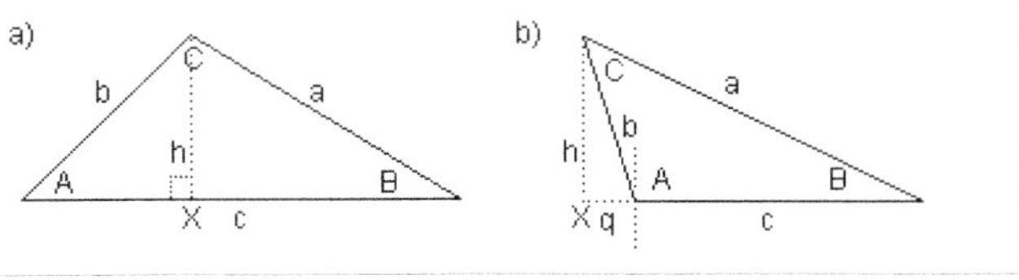

Angle C is greater than 90° in obtuse triangle ABC (Figure 502a). Draw altitude CX perpendicular to side c. Then

(1*a*) *in right triangle ACX* $h = b \sin A$

(1*b*) *in right triangle XCB* $h = a \sin B$

(1*c*) $\therefore \ b \sin A = a \sin B \quad \Rightarrow \quad \dfrac{\sin A}{a} = \dfrac{\sin B}{b} \quad or \quad \dfrac{b}{\sin B} = \dfrac{a}{\sin A}$

Now let angle C be less than π/2 and angle A greater than π/2 (Figure 502b). If you draw altitude CX perpendicular to side c the altitude falls outside of the triangle so that in the small triangle the angle is π–A.

$$(2a)\quad h = b\sin(\pi - A) = a\sin B \Rightarrow \quad \frac{\sin(\pi - A)}{a} = \frac{\sin B}{b}$$

however

$$(2b)\quad \sin(\pi - A) = \sin\pi\cos A - \cos\pi\sin A \quad (\textit{from Section 4.1})$$
$$= 0\times\cos A - (-1)\times\sin A = \sin A$$

$$(2c)\quad \textit{and so} \quad \frac{\sin A}{a} = \frac{\sin B}{b}$$

Then by the same reasoning

$$(2d)\quad c\sin A = a\sin C \quad \Rightarrow \quad \frac{\sin A}{a} = \frac{\sin C}{c}$$

$$(2e)\quad b\sin C = c\sin B \quad \Rightarrow \quad \frac{\sin C}{c} = \frac{\sin B}{b}$$

Combining 2c, 2d, 2e we get the Law of Sines.

$$(3a)\quad \frac{a}{\sin A} = \frac{b}{\sin B} = \frac{c}{\sin C} \qquad (3b)\quad \frac{\sin A}{a} = \frac{\sin B}{b} = \frac{\sin C}{c}$$

Problem 501. In a right triangle the hypotenuse c = 15 inches, and the sine of angle A, is sin A = 2/5. Find a, the side opposite A, and find b, the third side.

Problem 502. In a right triangle B = 55° 30', and b = 6.05. Find c and a.

Problem 503. The top of a ladder 50 feet long rests against a building 43 feet from the ground. At what angle does the ladder slope, and what is the distance of its foot from the wall?

Problem 504. If you know two sides and the angle opposite one of them, then you can almost determine the angle opposite the other one of them. For instance, if side a = 25, side b = 15, and angle A = 40°, then the law of sines says (sin 40°)/25 = (sin B)/15. Solving for sin B gives sin B = 15 (sin 40°)/25 = 0.38567. Now, the arcsin of 0.38567 = 22.686°.

5.2 The Law of Cosines

Figure 502 Obtuse Triangles

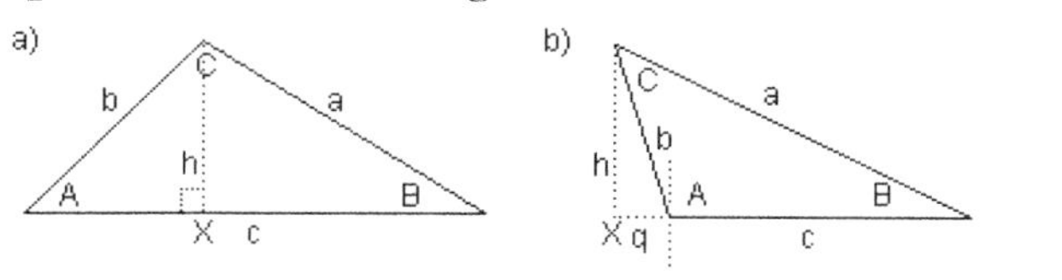

Focus on angle B in triangle ACX (Figure 502a):

$(4a)$ *in right triangle ACX* $b^2 = h^2 + AX^2$

$(4b)$ *and* $h = a\sin B,\quad AX = c - a\cos B$

$(4c)$ $\therefore\ b^2 = h^2 + AX^2 = (a\sin B)^2 + (c - a\cos B)^2$

$$b^2 = a^2\sin^2 B + c^2 - 2ac\cos B + a^2\cos^2 B$$

$$b^2 = a^2(\sin^2 B + \cos^2 B) + c^2 - 2ac\cos B$$

$$b^2 = a^2 + c^2 - 2ac\cos B$$

Focus on angle A in triangle ABC in Figure 502b:

(5) $\cos(\pi - \mathrm{A}) = \cos\pi\cos\mathrm{A} + \sin\pi\sin\mathrm{A}$ (Section 8.1)

$$= 1\times\cos\mathrm{A} + 0\times\sin\mathrm{A} = \cos\mathrm{A}$$

$(6a)$ *in right triangle ACX* $b^2 = h^2 + q^2$

$(6b)$ $a^2 = h^2 + (c+q)^2 = b^2 - q^2 + c^2 + 2qc + q^2$

$(6c)$ $a^2 = b^2 + c^2 + 2qc$

$(6d)$ $a^2 = b^2 + c^2 - 2bc\dfrac{q}{b} = b^2 + c^2 - 2bc\cos(\pi - A)$

$(6e)$ $a^2 = b^2 + c^2 - 2bc\cos A$

$(6f)$ *by analogy* $c^2 = a^2 + b^2 - 2ab\cos C$

The cosine law is valid for all interior angles of a triangle. If angle B=90° cos B=0 and the law becomes Pythagora's Theorem. The cosine law is written three ways (4c, 6e, 6f).

Problem 505. b = 2.25 meters and cos A = 0.15. Find a and c.
Problem 506. b = 12 feet and cos B = 1/3. Find c and a.
Problem 507. b = 6.4, c = 7.8. Find A and a.
Problem 508. A = 23° 15', c = 12.15. Find a and b.

5.3 Area of a Triangle, Two Ways

The area of a triangle is ½ the area of a parallelogram with the same base and altitude (Figure 503). If side c is the base and h is the altitude then

Figure 503

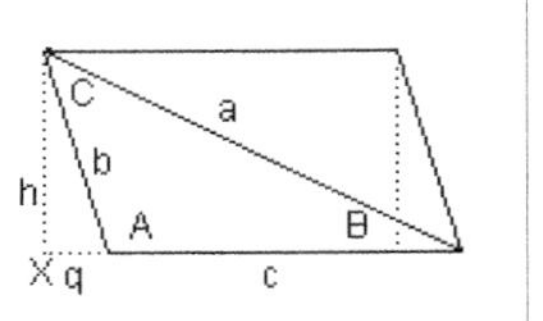

The area of a triangle is half the product of ANY two sides times the sine of the included angle.
(10) $\text{Area} = \frac{1}{2}\text{ch} = \frac{1}{2}\text{cb}\sin(\pi - \text{A}) = \frac{1}{2}\text{cb}\sin\text{A}$

Another way to compute the area of a triangle is via Heron's formula which gives the area in terms of the three sides of the triangle.

$$(11)\quad Area = \sqrt{s(s-a)(s-b)(s-c)} \quad where\ s = \frac{a+b+c}{2}$$

Here is one way to derive Heron's formula.

$$(12)\quad \cos A = \frac{b^2+c^2-a^2}{2bc}$$

$$(13)\quad \sin^2 A = 1-\cos^2 A = 1-\left(\frac{b^2+c^2-a^2}{2bc}\right)^2 = \frac{4b^2c^2-(b^2+c^2-a^2)^2}{4b^2c^2}$$

$$= \frac{(2bc+b^2+c^2-a^2)(2bc-b^2-c^2+a^2)}{4b^2c^2}$$

$$= \frac{(a+b+c)(b+c-a)(c+a-b)(a+b-c)}{4b^2c^2}$$

$$(14)\quad \text{Area} = \frac{1}{2}\text{ch} = \frac{1}{2}\text{cb}\sin\text{A} = \frac{1}{2}\text{cb}\sqrt{\frac{(\text{a}+\text{b}+\text{c})(\text{b}+\text{c}-\text{a})(\text{c}+\text{a}-\text{b})(\text{a}+\text{b}-\text{c})}{4\text{b}^2\text{c}^2}}$$

$$= \sqrt{\frac{(\text{a}+\text{b}+\text{c})(\text{b}+\text{c}-\text{a})(\text{c}+\text{a}-\text{b})(\text{a}+\text{b}-\text{c})}{2^4}}$$

$$= \sqrt{\text{s}(\text{s}-\text{a})(\text{s}-\text{b})(\text{s}-\text{c})} \quad \text{from 11}$$

5.4 Solving Obtuse Triangles

Two angles and a side Given two angles and the side opposite one of them, the side opposite the other one of them can be determined. For example, if angle u = 30°, angle v = 45°, and side U = 16, then apply the law of sines.

$$(1)\quad \frac{\sin 30}{U}=\frac{\sin 45^\circ}{V} \rightarrow \frac{\sin 30^\circ}{16}=\frac{\sin 45^\circ}{V} \rightarrow V=22.63$$

Three sides Given three sides a, b, c of a triangle use the law of cosines to find any angle.

$(2a)\quad c^2=a^2+b^2-2ab\cos C$

$(2b)\quad \textit{if}\ \ a=5,\ b=6,\ c=9\ \ \textit{then}\ 81=25+36-60\cos C$

$$(2c)\quad \cos C=\frac{81-25-36}{60}=-\frac{20}{60}=-0.333..$$

$(2d)\quad C=\arccos(-0.33..)=109.5^\circ$

Two sides and an angle Given one angle and the two adjacent sides find the opposite side.

$(3a)\quad c^2=a^2+b^2-2ab\cos C$

$(3b)\quad \textit{if}\ \ a=5,\ b=6,\ C=65^\circ\ \textit{then}\ c^2=25+36-60\cos 65^\circ$

$(3c)\quad c^2=61-60\times 0.4226=61-25.36=35.64$

$(3d)\quad c=5.97$

Two sides and an included angle Given two sides U=20, V=15 and the angle u=40° opposite side U find the angle v opposite side V.

$$(4)\quad \frac{\sin u}{U}=\frac{\sin v}{V} \rightarrow \frac{\sin 40^\circ}{20}=\frac{\sin v}{15} \rightarrow v=28.82^\circ$$

Note: 28.82° may not be the correct answer. There are two angles between 0 and 180° with the same sine value. The second angle is the supplement of the first or 180 – 28.82 = 157.18°. Knowing two sides and the angle opposite one of them may not be enough to determine the triangle. There is no deterministic *side-angle-side* congruence theorem in geometry.

6 Trigonometry and Circles

Trigonometric functions can use any angle, whereas right triangles are limited to one 90° angle and two acute angles ranging from 0° to 90°.

In order to work with the trigonometric functions of any angle define the trigonometric sine and cosine functions by using a circle of radius 1, i.e. the unit circle. The other trigonometric functions are defined as combinations of the cosine and sine.

Begin by constructing an angle in standard position, where the initial side is placed on the positive x axis. Then the terminal side of the angle u will intersect the unit circle at some point x, y (Figure 601). The cosine of the angle is the x coordinate of the point divided by the unit radius so that the cosine numerically equals the x coordinate. Similarly the sine equals the y coordinate (Figure 602).

Figure 601

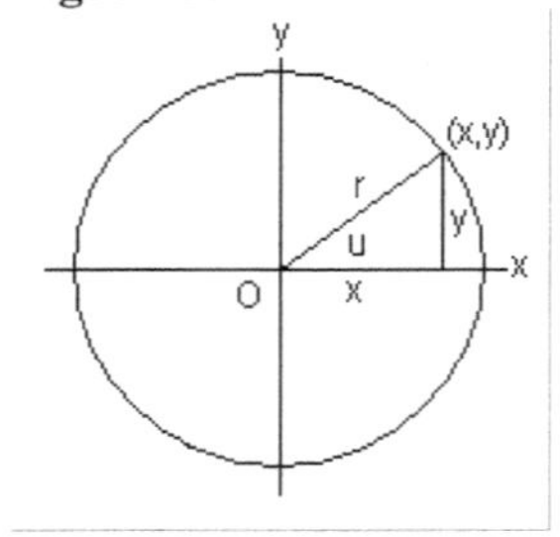

Figure 602

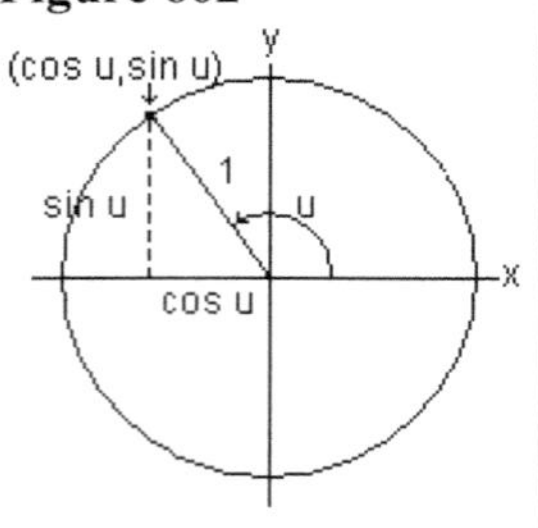

Observe that the values of the trigonometric functions depend only on the intersection of the terminal side and the unit circle. In this way every angle is associated with a point on the unit circle, and every point on the unit circle is associated with a multiplicity of angles $\theta+2\pi n$.

And so the trigonometric functions are defined for acute angles in two ways as ratios in a right triangle and as points on the unit circle. The two definitions should agree. Consider the following argument.

A right triangle is formed by dropping down a line from the (x,y) point of intersection to the x axis (Figure 601). The hypotenuse is the length of the radius, which is 1 on this unit circle. The initial side of the triangle is the value x, and the terminal side of the triangle is the value y. Consequently sine and cosine are as follows.

$$(1)\quad \sin u = \frac{y}{1} = y \quad \rightarrow \quad \cos u = \frac{x}{1} = x$$

Clearly the two ways of defining the trigonometric functions agree for acute angles.

In right triangles the trigonometric functions are always positive. Increasing the range of angles beyond 90° produces values of the sine and cosine functions that are negative as well as positive (Figure 603). The sine function takes the *sign of y* and the cosine function takes the *sign of x*.

Figure 603

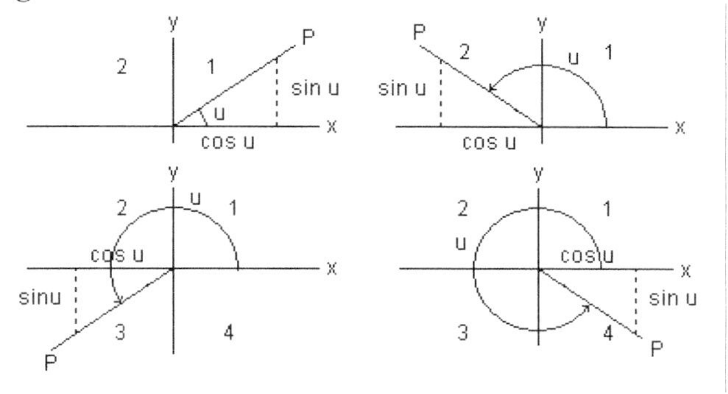

$$(2a)\quad \sin(u + 2\pi n) = \sin u \qquad (2b)\quad \cos(u + 2\pi n) = \cos u$$

Sine and cosine are sine and cosine of complementary angles.

$$(3a)\quad \sin u = \cos(\pi / 2 - u) \qquad (3b)\quad \cos u = \sin(\pi / 2 - u)$$

Consider sine and cosine as functions. A function *f(x)* is an *odd* function if for any *x*, $f(-x) = -f(x)$. A function *f(x)* an *even* function if for any number *x*, $f(-x) = f(x)$. Sine is an odd function, and cosine is an even function (Fig 203 page 88). Note: Most functions are neither odd nor even functions.

$$(4a)\quad \sin(-u) = -\sin u \qquad (4b)\quad \cos(-u) = \cos u$$

Geometry

Similar triangles and the Pythagorean theorem are used extensively in trigonometry. A right triangle has one 90 degree angle.

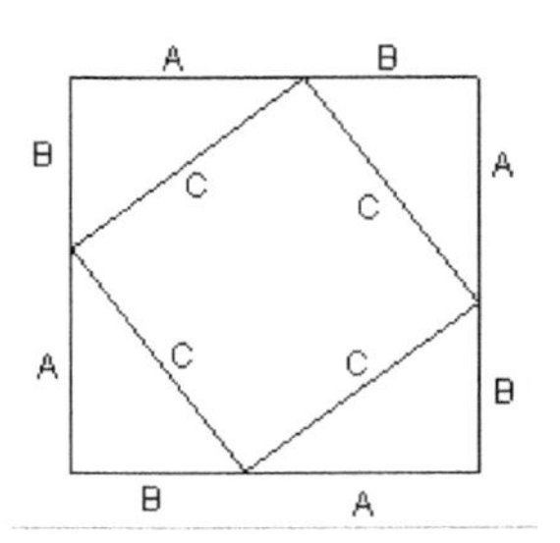

The Pythagorean theorem Draw right triangle ABC. Draw square with sides A+B in length. Mark distance B on each side. Draw three more C lines to create four identical triangles. Calculate area of A+B square two ways.

$$Area = (A+B)^2 = C^2 + 4\times\frac{AB}{2}$$

$$A^2 + 2AB + B^2 = C^2 + 2AB$$

$$A^2 + B^2 = C^2$$

This demonstrates the Pythagorean theorem, which states that the square of the hypotenuse C is the sum of the squares of the two sides A and B, so that $C^2=A^2+B^2$

Similar triangles Two triangles ABC and DEF are similar if
(1) corresponding angles are equal, or,
(2) their sides are proportional so that the ratios of the three corresponding sides are equal:

$$\frac{A}{D} = \frac{B}{E} = \frac{C}{F}$$

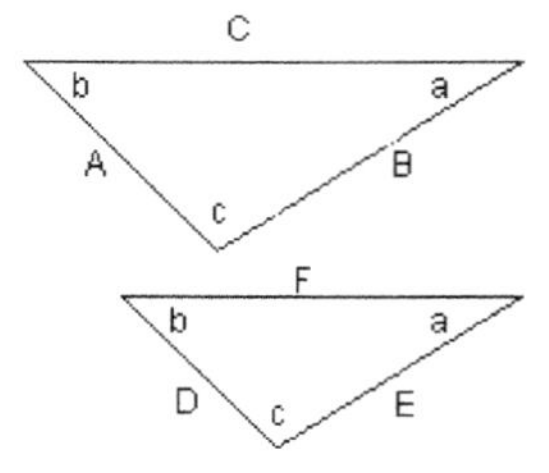

The smaller of the two similar triangles is part of the larger. For example, DEF is part of ABC, while sharing angle a. Observe that opposite sides A and D are parallel lines. When the similar triangles are separated the similarity is not obvious.

Any triangle is the sum of two right triangles In a triangle the sum of the angles is 180° (degrees). Divide the triangle by dropping a perpendicular from angle c to the opposite side C. Angle c can be any size.

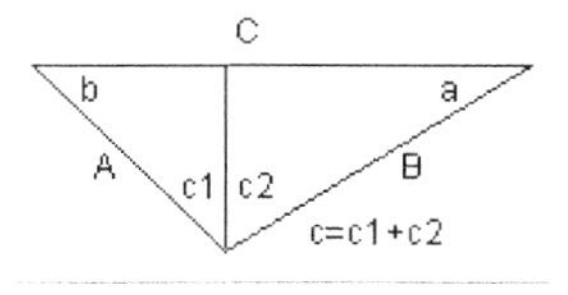

The value of π The value of π the first ten places of decimals is 3.1415926536. Useful approximations are

$$\frac{22}{7} = 3.142857143 \qquad \frac{355}{113} = 3.1415920$$

Projections Draw perpendicular line PB from point P to point B on the y axis. Draw perpendicular line PA from point P to point A on the x axis. Line OB is the projection of OP onto the y axis, and line OA is the projection of OP onto the x axis.

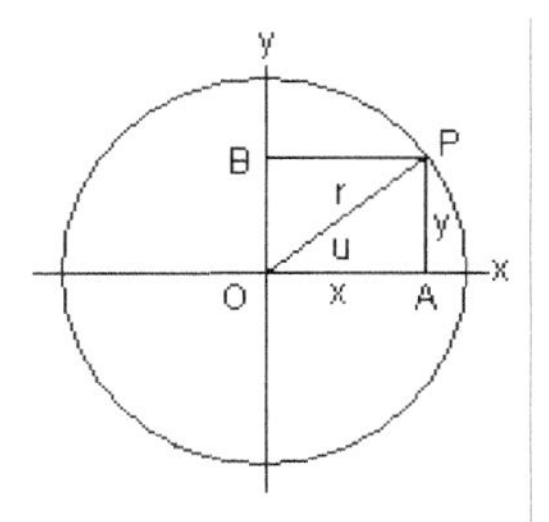

Pythagorean triples A triple is a set of three numbers that satisfy the Pythagorean theorem. For example

$$(x,y,r) \quad (3,\ 4,\ 5) \quad (1,\ 1,\ \sqrt{2}) \quad (1,\ \sqrt{3},\ 2) \quad (5,\ 12,\ 13) \quad (9,\ 40,\ 41)$$

Angle in standard position An angle is in standard position when the vertex is at the origin O of the x,y plane, and the initial side is on the positive x axis.

Coterminal Angles These are angles in standard position, which have coincident terminal sides.

Parallel Lines and their angles

$$u + y = \frac{\pi}{2} \quad \rightarrow \quad z = \frac{\pi}{2} - y = \frac{\pi}{2} - \frac{\pi}{2} + u \quad \rightarrow \quad z = u$$

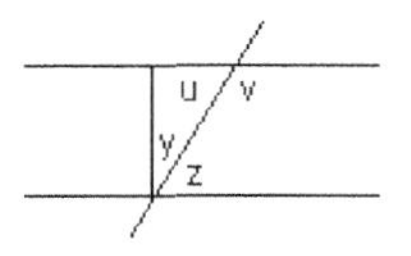

7 Inverse Trigonometric Functions

The basic idea is that f^{-1} "undoes" what f does, and vice versa.

(1*a*) $f^{-1}f(x) = x$ *for all x in the domain of f(x)*

(1*a*) $f(f^{-1}(y)) = y$ *for all y in the domain of f(y)*

The numbers in a *domain* are *inputs* to a function f(x), and the numbers in the *range* are the outputs (see sidebar).

(2) 5 *is an input to* $\sin x$, *and* $\sin 5 = 0.0872$, *which is the output*

For inputs $-\pi/2 \leq x \leq \pi/2$ the outputs are $-1 \leq \sin x \leq 1$.

Define the *inverse sine* function $u = \sin^{-1} x$ (also written as $u = \arcsin x$) whose domain is the interval (−1, 1) and whose range is the interval $-\pi/2$ to $\pi/2$.

(3*a*) $u = \sin^{-1}(\sin u)$ *for* $-\pi/2 \leq u \leq \pi/2$

u is the angle whose sine is $\sin u$

(3*b*) $x = \sin(\sin^{-1} x)$ *for* $-1 \leq x \leq 1$

x is the sine of the angle whose sine is x

(4)

Inverse function	*Principle value range*
$u = \arcsin x$	$-\pi/2 \leq u \leq \pi/2$
$u = \arccos x$	$0 \leq u \leq \pi$
$u = \arctan x$	$-\pi/2 \leq u \leq \pi/2$
$u = \operatorname{arccot} x$	$0 \leq u \leq \pi$
$u = \operatorname{arcsec} x$	$0 \leq u \leq \pi,\ u \neq \pi/2$
$u = \operatorname{arecsc} x$	$-\pi/2 \leq u \leq \pi/2,\ u \neq 0$

Trigonometric functions of inverse trigonometric functions may seem strange if not mysterious. Consider these two items.

(5*a*) $C_n(\omega) = \cos(n \arccos \omega) = f(\omega)$ $|\omega| \leq 1$ *Chebyshev polynomial* $f(\omega)$

(5*b*) $y = \cot\left(\arccos\left(\frac{x-1}{x+1}\right)\right) = f(x)$

A typical right angle triangle problem – find y=f(x):

y is the trig function of some angle u (6a)
angle u is some inverse trig function of some f(x) (6b)
thus some trig function such as cos u = f(x) (6b)
interpret f(x) as the ratio of two polynomials N(x)/D(x)=a/c (6b)
N(x) and D(x) are 2 sides of a right triangle. Calculate side 3 E(x) (6c)
y is some ratio of N(x), D(x), and E(x) (6d)

For example

$$\text{(6a)}\quad y = \cot\left(\arccos\left(\frac{x-1}{x+1}\right)\right) = \cot u = \frac{\cos u}{\sin u}$$

$$\text{(6b)}\quad u = \arccos\left(\frac{x-1}{x+1}\right) \;\rightarrow\; \cos u = \frac{x-1}{x+1} = \frac{a}{c}$$

$$\text{(6c)}\quad b^2 = c^2 - a^2 = (x+1)^2 - (x-1)^2 = 4x \;\rightarrow\; b = 2\sqrt{x} \;\rightarrow\; \sin u = b/c$$

$$\text{(6d)}\quad y = \cot u = \frac{\cos u}{\sin u} = \frac{a/c}{b/c} = \frac{a}{b} = \frac{x-1}{2\sqrt{x}}$$

Domain, Range, One-to-One

A function is a rule that assigns a single object y from one set (the **range**) to each object x from another set (the **domain**). We can write that rule as $y = f(x)$, where f is the function. There is a simple *vertical rule* for determining whether a rule $y = f(x)$ is a function: f is a function if and only if every vertical line intersects the graph of $y = f(x)$ in the xy-coordinate plane at most once.

A function f is *one-to-one* if it assigns distinct values of y to distinct values of x. In other words, if $6x_1 = x_2$ then $6f(x_1) = f(x_2)$. Equivalently, f is one-to-one if $f(x1) = f(x2)$ implies $x_1 = x_2$. There is a simple *horizontal rule* for determining whether a function $y = f(x)$ is one-to-one: f is one-to-one if and only if every horizontal line intersects the graph of $y = f(x)$ in the xy-coordinate plane at most once.

If a function f is one-to-one on its domain, then f has an *inverse function*, denoted by f^{-1}, such that $y = f(x)$ if and only if $f^{-1}(y) = x$. The domain of $f-1$ is the range of f.

---

8 Trigonometric Functions of Two or More Angles

There are a large number of trigonometric identities. So many that we do not even try to memorize them, because there is a straightforward method for deriving what we may need at any moment. The basis of the method is Euler's famous identity (8.3). However start by deriving sum and difference identities based on a geometric figure.

8.1 Sums and Differences of Two Angles

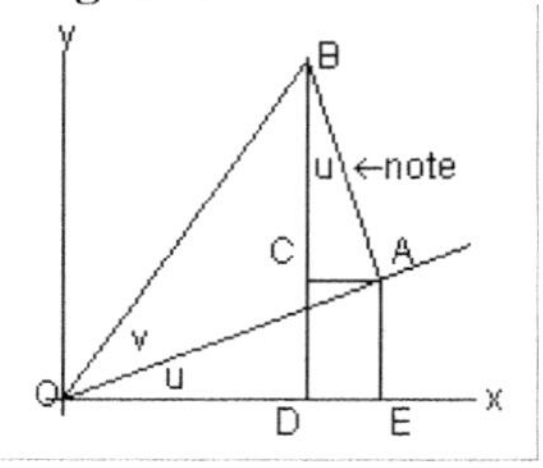

We construct figures to use as a basis for derivations. Place angle u in standard position (Figure 801). Place angle v with its vertex at the origin and with its initial side coincident with the terminal side of angle u. From any point B on the terminal side of v draw BD perpendicular to the x axis, and BA perpendicular to OA. Draw AE perpendicular to the x axis. Draw AC perpendicular to BD. Now we can say

Figure 801

$$(1)\ \sin(u+v) = \frac{DC+CB}{OB} = \frac{AE}{OB} + \frac{CB}{OB} = \frac{AE}{OA}\frac{OA}{OB} + \frac{CB}{AB}\frac{AB}{OB}$$

$$= \sin u \cos v + \cos u \sin v$$

$$(2)\ \cos(u+v) = \frac{OD}{OB} = \frac{OE-DE}{OB} = \frac{OE}{OB} - \frac{CA}{OB} = \frac{OE}{OA}\frac{OA}{OB} - \frac{CA}{AB}\frac{AB}{OB}$$

$$= \cos u \cos v - \sin u \sin v$$

$$(3a)\quad y = \tan(u+v) = \frac{\sin(u+v)}{\cos(u+v)} = \frac{\sin u \cos v + \cos u \sin v}{\cos u \cos v - \sin u \sin v}$$

$$(3b)\quad \text{divide by } \cos v \ \rightarrow\ y = \frac{\sin u + \cos u \tan v}{\cos u - \sin u \tan v}$$

$$(3c)\quad \text{divide by } \cos u \ \rightarrow\ y = \tan(u+v) = \frac{\tan u + \cos u \tan v}{1 - \tan u \tan v}$$

8.2 Double Angles and Half Angles

(1) $\sin(u+v) = \sin u \cos v + \cos u \sin v$

(2) $\cos(u+v) = \cos u \cos v - \sin u \sin v$

(3) $\tan(u+v) = \dfrac{\tan u + \tan v}{1 - \tan u \tan v}$

(4) $\sin(u-v) = \sin u \cos v - \cos u \sin v$

(5) $\cos(u-v) = \cos u \cos v - \sin u \sin v$

(6) $\tan(u-v) = \dfrac{\tan u - \tan v}{1 + \tan u \tan v}$

Equations 1 and 2 produce double angle equations.

$\sin 2u = \sin(u+u) = \sin u \cos u + \cos u \sin u$

(7) $\sin 2u = 2 \sin u \cos u$

(8*a*) $\cos 2u = \cos(u+u) = \cos u \cos u - \sin u \sin u = \cos^2 u - \sin^2 u$

(8*b*) $\cos 2u = 2\cos^2 u - 1$

(8*c*) $\cos 2u = 1 - 2\sin^2 u$

The half angle equations follow immediately.

Substitute $v = 2u$ *in equations* 8

$\cos v = 2\cos^2 \dfrac{v}{2} - 1$ $\qquad$ $\cos v = 1 - 2\sin^2 \dfrac{v}{2}$

(9*a*) $\cos \dfrac{v}{2} = \pm\sqrt{\dfrac{1+\cos v}{2}}$ $\qquad$ (9*b*) $\sin \dfrac{v}{2} = \pm\sqrt{\dfrac{1-\cos v}{2}}$

(10*a*) $\tan \dfrac{v}{2} = \pm\sqrt{\dfrac{1-\cos v}{1+\cos v}}$

$$\tan \frac{v}{2} = \pm\sqrt{\frac{1-\cos v}{1+\cos v} \times \frac{1+\cos v}{1+\cos v}} = \pm\sqrt{\frac{1-\cos^2 v}{(1+\cos v)^2}}$$

(10*b*) $\tan \dfrac{v}{2} = \dfrac{\sin v}{1+\cos v}$

(10*c*) $\tan \dfrac{v}{2} = \dfrac{1-\cos v}{\sin v}$

Problem 801 Show that $\cos\left(\frac{\pi}{2}-u\right)=\sin u$ *and* $\sin\left(\frac{\pi}{2}-u\right)=\cos u$

Problem 802 Derive equation 10c.

Problem 803 If cos u = √3/2 find cos 2u.

Problem 804 If cos u = √3/2 find sin 2u.

Problem 805 Derive an equation for tan 2u.

Problem 806 Derive an equation for tan(u/2).

Problem 807 Show that $\cos(u+2\pi)=\cos u$ $\quad \sin(u+2\pi)=\sin u$

Problem 808 Show that $\cos(-u)=\cos(u)$ $\quad \sin(-u)=-\sin u$

Products to sums

(21) $2\cos u\cos v=\cos(u+v)+\cos(u-v)$

(22) $-2\sin u\sin v=\cos(u+v)-\cos(u-v)$

(23) $2\sin u\cos v=\sin(u+v)+\sin(u-v)$

(24) $2\cos u\sin v=\sin(u+v)-\sin(u-v)$

Problem 809 Derive equation 21
Problem 810 Derive equation 22
Problem 811 Derive equation 23
Problem 812 Derive equation 24

Sums to products

(25) $\sin u+\sin v=2\sin\frac{1}{2}(u+v)\cos\frac{1}{2}(u-v)$

(26) $\sin u-\sin v=2\cos\frac{1}{2}(u+v)\sin\frac{1}{2}(u-v)$

(27) $\cos u+\cos v=2\cos\frac{1}{2}(u+v)\cos\frac{1}{2}(u-v)$

(28) $\cos u-\cos v=-2\sin\frac{1}{2}(u+v)\sin\frac{1}{2}(u-v)$

Problem 813 Derive equation 25
Problem 814 Derive equation 26
Problem 815 Derive equation 27
Problem 816 Derive equation 28

8.3 Derivations Using Euler's Identity

(100) $e^{\pm iu} = \cos u \pm i \sin u \quad \rightarrow \quad i = \sqrt{-1} \quad and \quad i^2 = -1$

Sums and differences of two angles

$$e^{i(u+v)} = e^{iu} e^{iv}$$

$$\cos(u+v) + i\sin(u+v) = [\cos u + i \sin u][\cos v + i \sin v]$$

$$= [\cos u \cos v - \sin u \sin v] + i[\sin u \cos v + \cos u \sin v]$$

(101) $\sin(u+v) = \sin u \cos v + \cos u \sin v$

(102) $\cos(u+v) = \cos u \cos v - \sin u \sin v$

Problem 817 Use Euler's identity. Show that

(103) $\sin(u-v) = \sin u \cos v - \cos u \sin v$

(104) $\cos(u-v) = \cos u \cos v + \sin u \sin v$

Problem 818 Show that $\tan\ (u+v) = \dfrac{\tan u + \tan v}{1 - \tan u \tan v}$

Problem 819 Show that $\tan\ (u-v) = \dfrac{\tan u - \tan v}{1 + \tan u \tan v}$

Identities relating angles and their complements Observe how the left and right sides are processed.

$$e^{i(-u)} e^{i\pi/2} = e^{-iu} e^{i\pi/2}$$

$$e^{i(\pi/2-u)} = e^{-iu} e^{i\pi/2}$$

$$\cos\left(\frac{\pi}{2} - u\right) + i\sin\left(\frac{\pi}{2} - u\right) = [\cos u - i \sin u][\cos\left(\frac{\pi}{2}\right) + i\sin\left(\frac{\pi}{2}\right)]$$

$$= [\cos u - i \sin u][0 + i1]$$

(105*a*) $\cos\left(\dfrac{\pi}{2} - u\right) + i\sin\left(\dfrac{\pi}{2} - u\right) = \sin u + i \cos u$

In equation 105a equate real and imaginary to real and imaginary.

(105*b*) $\cos\left(\dfrac{\pi}{2} - u\right) = \sin u$ (106) $\sin\left(\dfrac{\pi}{2} - u\right) = \cos u$

Problem 820 Derive equations $\cot\left(\frac{\pi}{2}-u\right)=\tan u \quad \tan\left(\frac{\pi}{2}-u\right)=\cot u$

Problem 821 Derive equations $\csc\left(\frac{\pi}{2}-u\right)=\sec u \quad \sec\left(\frac{\pi}{2}-u\right)=\csc u$

Problem 822 Show that $\tan(2u)=\frac{2\tan u}{1-\tan^2 u}$

Functions are periodic.

$$e^{iu}e^{i2\pi}=e^{iu}e^{i2\pi} \quad\rightarrow\quad e^{i(u+2\pi)}=e^{iu}e^{i2\pi}$$

$$\begin{aligned}\cos(u+2\pi)+i\sin(u+2\pi)&=[\cos u+i\sin u][\cos 2\pi+i\sin 2\pi]\\&=[\cos u+i\sin u][1+i0]\\&=\cos u+i\sin u\end{aligned}$$

(109) $\cos(u+2\pi)=\cos u$ (110) $\sin(u+2\pi)=\sin u$

Symmetry

$$e^{i(-u)}=e^{-iu}$$

$$\cos(-u)+i\sin(-u)=[\cos(u)-i\sin(u)]$$

(111) $\cos(-u)=\cos(u)$ (112) $\sin(-u)=-\sin u$

Half-angles

$$e^{iu/2}e^{iu/2}=e^{iu/2}e^{iu/2} \quad\rightarrow\quad e^{iu}=e^{iu/2}e^{iu/2}$$

$$\begin{aligned}\cos u+i\sin u&=[\cos\frac{u}{2}+i\sin\frac{u}{2}][\cos\frac{u}{2}+i\sin\frac{u}{2}]\\&=[\cos\frac{u}{2}\cos\frac{u}{2}-\sin\frac{u}{2}\sin\frac{u}{2}]+i[\sin\frac{u}{2}\cos\frac{u}{2}+\cos\frac{u}{2}\sin\frac{u}{2}]\end{aligned}$$

$$\cos u=\cos^2\frac{u}{2}-\sin^2\frac{u}{2}=2\cos^2\frac{u}{2}-1=1-2\sin^2\frac{u}{2}$$

(113) $\cos\frac{u}{2}=\sqrt{\frac{1+\cos u}{2}}$ (114) $\sin\frac{u}{2}=\sqrt{\frac{1-\cos u}{2}}$

Problem 823 Show that $\tan\frac{u}{2}=\pm\sqrt{\frac{1-\cos u}{1+\cos u}}$

Problem 824 Show that $\tan\frac{u}{2}=\frac{1-\cos u}{\sin u}=\frac{\sin u}{1+\cos u}$

---

9 Differentiation

Differentiation of a function f(x) means finding the relationship of a change in f(x), written as df(x) *dee f of x*, and a change in x written as dx *dee x*. The term *df/dx* is referred to as the derivative of *f(x)*. What follows are very informal processes avoiding limits and other formalities in order to focus on the derivatives.

Consider the function $y = \sin u$ *and find* $\dfrac{dy}{du} = \dfrac{d\sin u}{du}$

An increment in y is marked a *dy*, and increment in u is marked as *du*.

(1a) $y + dy = \sin(u + du)$

(1b) $dy = \sin(u + du) - y = \sin(u + du) - \sin u$

(eqn 26 p110) $\sin u - \sin v = 2\cos\frac{1}{2}(u + v)\sin\frac{1}{2}(u - v)$

(1c) $dy = 2\cos\frac{1}{2}(u + du + u)\sin\frac{1}{2}(u + du - u)$

(1d) $dy = 2\cos(u + \frac{1}{2}du)\sin(\frac{1}{2}du)$

(1e) if $u + \frac{1}{2}du \approx u$ and $\sin(\frac{1}{2}du) \approx \frac{1}{2}du$ then

(1f) $dy = 2\cos u \times \frac{1}{2}du = \cos u\, du$

$$(2)\quad \frac{dy}{du} = \frac{d\sin u}{du} = \cos u$$

Consider the function $y = \cos u$ *and find* $\dfrac{dy}{du} = \dfrac{d\cos u}{du}$

$$(3a)\quad y = \cos u = \sin\left(\frac{\pi}{2} - u\right)$$

$$(3b)\quad d(constant) = 0 \qquad \text{constants do not change}$$

$$(3c)\quad d\left(\frac{\pi}{2} - u\right) = d\frac{\pi}{2} - du = 0 - du = -du$$

$$(3d)\quad dy = d\sin\left(\frac{\pi}{2} - u\right) = \cos\left(\frac{\pi}{2} - u\right) \times d\left(\frac{\pi}{2} - u\right) = \cos\left(\frac{\pi}{2} - u\right) \times -du = -\sin u\, du$$

$$(4)\quad \frac{dy}{du} = \frac{d\cos u}{du} = -\sin u$$

10 Integration

The integration process is the reverse of the differentiation process. Formulating many physical problems mathematically produces derived functions such as Newton's force equation $f = ma = m\ d^2x/dt^2$. Consequently knowing the derivative requires finding the function. For example

$(5a)$ *the long s* $\int$ *means "the sum of"*

$(5b)$ $\int dy$ *means the sum of the dy so that* $\int dy = y$

Then

$$(6a)\quad \frac{dy}{du} = \cos u \;\rightarrow\; dy = \cos u\, du$$

$$(6b)\quad \int dy = \int \cos u\, du \;\rightarrow\; y = \sin u + constant\ C$$

The integral of the cosine is known because the derivative of the sine is known. In a sense this is cheating. In fact all integration processes depend on a table of $f(x)$, their derivatives $df(x)/dx$, as well as integration techniques such as integration by parts, partial fractions, substitution, change of variable. Furthermore there are extensive tables of integrals available.

In any calculus book the formula for integration by parts is derived.

$$(7)\quad \int u\, dv = uv - \int v\, du$$

Integrate

$$(8a)\quad \int x \sin x\, dx$$

$$(8b)\quad if\ \ u = x \quad \frac{dv}{dx} = \sin x \ \ then\ \ du = dx \quad v = -\cos x$$

$$(8c)\quad \int x \sin x\, dx = uv - \int v du = -x\cos x - \int(-\cos x)\, dx$$

$$(8d)\quad \int x \sin x\, dx = -x\cos x + \sin x + C$$

11 Hyperbolic Functions

Hyperbolic functions are derived from the hyperbola, whereas the trigonometric functions are derived from the circle. Hyperbolic functions and equations parallel the trigonometric functions and equations.

The formula for area dA of any sector OPQ (Figure 1101) is from geometry,

$$(1a)\quad dA = \frac{1}{2} r^2 du \quad \to \quad du = \frac{1}{r^2} 2dA$$

Figure 1101

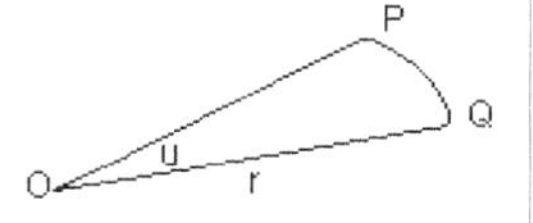

In rectangular coordinates part-of-a-circle area dA is calculated as follows.

$$(2a)\quad r^2 = x^2 + y^2 \quad and \quad \tan u = \frac{y}{x}$$

$$(2b)\quad du = d\left(arctan \frac{y}{x} \right) = \frac{xdy - ydx}{x^2 + y^2}$$

$$(2c)\quad dA = \frac{1}{2} r^2 du = \frac{1}{2}\left(x^2 + y^2\right)\left(\frac{xdy - ydx}{x^2 + y^2} \right)$$

$$(2d)\quad dA = \frac{1}{2}(xdy - ydx)$$

Angle u is what is of immediate interest, because it leads to *cos u*.

$$(3)\quad if \;\; r = 1 \;\; then \;\; y = \sqrt{1 - x^2} \quad and \quad dy = -\frac{x}{\sqrt{1-x^2}} dx$$

$$(4a)\quad if \;\; r = 1 \;\; then \;\; du = 2dA = xdy - ydx \quad from \;\; 1a$$

$$(4b)\quad du = x\left(-\frac{x}{\sqrt{1-x^2}} dx \right) - \sqrt{1-x^2}\, dx = \frac{-dx}{\sqrt{1-x^2}}$$

$$(4c)\quad u = \int_1^x \frac{-dx}{\sqrt{1-x^2}} = \arccos x$$

$$(4d)\quad x = \cos u$$

The y side produces the sine.

$$(5)\quad y = \sqrt{1\text{-}x^2} = \sqrt{1\text{-} \cos^2 u} = \sin u$$

In rectangular coordinates part-of-a-hyperbola area dA is calculated as follows.

$$(6a) \quad r^2 = x^2 - y^2$$

$$(6b) \quad if \;\; r = 1 \;\; then \;\; y = \sqrt{x^2 - 1} \quad and \quad dy = \frac{x}{\sqrt{x^2 - 1}} dx$$

$$(7a) \quad if \;\; r = 1 \;\; then \;\; du = 2dA = xdy - ydx \qquad from \;\; 1a$$

$$(7b) \quad du = x\left(\frac{x}{\sqrt{x^2 - 1}} dx\right) - \sqrt{x^2 - 1}\, dx = \frac{dx}{\sqrt{x^2 - 1}}$$

$$(7c) \quad u = \int_1^x \frac{dx}{\sqrt{x^2 - 1}} = \ln\left(x + \sqrt{x^2 - 1}\right)$$

$$(7d) \quad e^u = x + \sqrt{x^2 - 1} \;\rightarrow\; e^u - x = \sqrt{x^2 - 1}$$

$$(7e) \quad e^{2u} - 2xe^u + x^2 = x^2 - 1 \;\rightarrow\; e^{2u} + 1 = 2xe^u \;\rightarrow\; e^u + \frac{1}{e^u} = 2x$$

$$(7f) \quad x = \frac{e^u + e^{-u}}{2} = \cosh u \qquad definition \;\; of \;\; \cosh u$$

$$(8a) \quad y = \sqrt{x^2 - 1} = \sqrt{\cosh^2 u - 1}$$

$$(8b) \quad y = \sqrt{\frac{(e^u + e^{-u})^2}{2^2} - 1} = \sqrt{\frac{e^{2u} + 2 + e^{-2u} - 4}{2^2}}$$

$$y = \sqrt{\frac{e^{2u} - 2 + e^{-2u}}{2^2}} = \frac{e^u - e^{-u}}{2} = \sinh u$$

Problem 1101 Show that $\cosh^2 u - \sinh^2 u = 1$

Problem 1102 Show that $u = \sinh^{-1} y = \ln\left(x + \sqrt{x^2 + 1}\right)$

Problem 1103 Show that $\sinh iu = i \sin u$

Problem 1104 Show that $\cosh iu = \cos u$

Problem 1104 Show that $\sin(x + iy) = \sin x \cosh y + i \cos x \sinh y$

Problem 1104 Show that $\cos(x + iy) = \cos x \cosh y - i \sin x \sinh y$

Problem 1104 Show that $\sinh(x + iy) = \sinh x \cos y + i \cosh x \sin y$

Problem 1104 Show that $\cosh(x + iy) = \cosh x \cos y + i \sinh x \sin y$

Appendix

A1 Absolute Value

Definition The absolute value of x, denoted as |x|, is defined as follows.

$$|x| = \begin{cases} x & if \quad x \geq 0 \\ -x & if \quad x \leq 0 \end{cases}$$

However the absolute value itself is *always* positive. $|x| \geq 0$

The *absolute value* of x on the real number line is the distance from 0 to x.

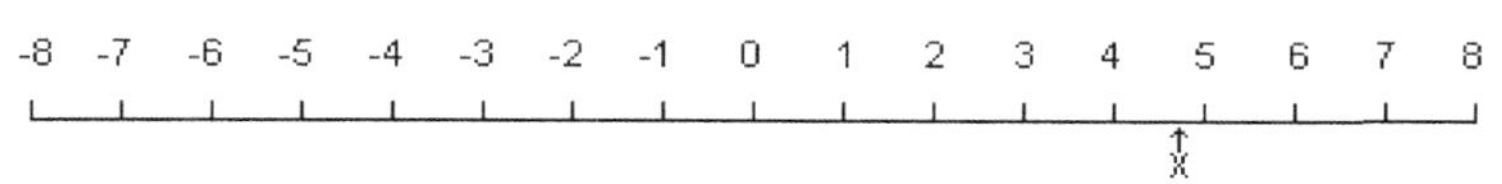

Since the absolute value of –q is q the solutions to the equation |x| = q are q and –q so that

$If \ |x| = 7, \ then \ x = 7 \ or \ x = -7$

$If \ |x| = q, \ then \ x = q \ or \ x = -q$

and

$If \ |x| = |y|, \ then \ x = y \ or \ x = -y$

Inequalities

$If \ q > 0 \ and \ |x| \leq q, \ then \ -q \leq x \leq q$

$If \ q > 0 \ and \ |x| \geq q, \ then \ x \leq -q \ and \ x \geq q$

Examples

$If \ |3x-5| = |28|, \ then \ 3x-5 = 28 \ or \ 3x-5 = -28 \ so \ that \ x = \frac{33}{3} = 11 \ or \ x = -\frac{23}{3}$

$If \ |2x-5| < 6, \ then \ -6 < 2x-5 < 6$

$Add \ 5 \ to \ each \ side \ -6+5 < 2x-5+5 < 6+5 \ \rightarrow \ -1 < 2x < 11 \ \rightarrow \ -\frac{1}{2} < x < \frac{11}{2}$

$If \ |2x-5| = |x-4|, \ then \ 2x-5 = x-4 \ \ or \ \ 2x-5 = -(x-4)$

$So \ that \ x = 1 \ or \ x = 3$

A2 Complex Numbers

The words complex and imaginary are potentially misleading, because complex numbers are not complicated and imaginary operators are not part of someone's imagination. Both words are labels: they are technical terms used to designate a class of numbers. A complex number z is represented by an ordered pair of real numbers x and y written as (x, y).

Multiplication by –1 and √–1 A number can be represented as a distance on a number line. We define steps to the right as positive so that distance AB=+4. Multiply +4 by –1 to get –4 that is the distance AC. Multiply AC by –1 to get back to AB. Clearly multiplication by –1 in effect *rotates* AB and AC by 180°.

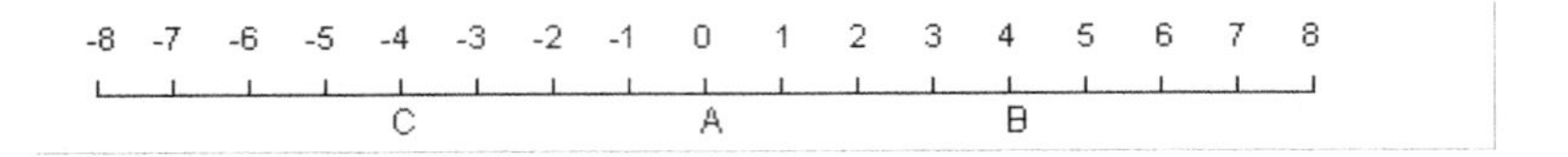

If +4 is multiplied by √–1 the result is 4√–1. Multiply 4√–1 by √–1 to get –4. Hence multiplication by √–1 two times rotates AB by 180°. And so multiplication by √–1 implements a 90° rotation of AB.

The world has agreed that numbers such as 4√–1 are *imaginary* numbers. To save writing √–1 is replaced by *i* in the mathematical literature.

Complex numbers The ordered pair (x_1, y_1) is a point in the (x, iy) plane that can be reached by starting from the origin, marching along the x-axis for a distance x_1, rotating $\pi/2$ radians, and marching parallel to the iy axis for distance y_1 (Figure A21a).

Working with ordered pairs (x, y) does not have much appeal, which is why the world adopted the well known alternative z=x+iy that is easier to work with.

In other words: taking our clue from the rotation operation we use i as a $\pi/2$ rotation operator. Then we say iy_1 is a vector we add to vector x_1 so that $z_1=x_1+iy_1$. This replaces the ordered pair (x_1, y_1). We say z is a complex number whose real part is x and whose imaginary part is y. Keep in mind that x and y are real numbers.

Figure A21 Complex numbers in Cartesian and polar coordinates

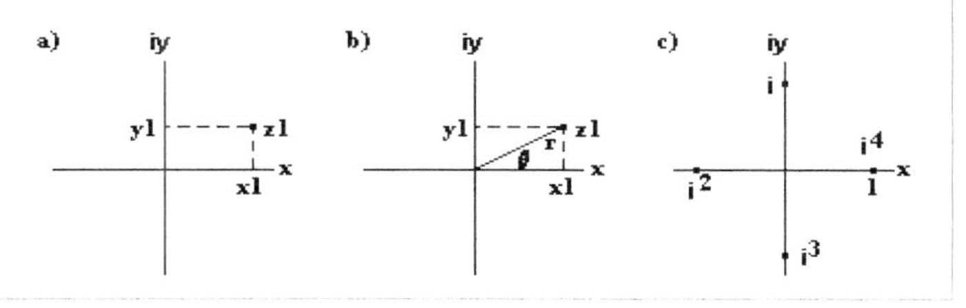

Polar coordinates: If r is the distance from the origin to the point z, then x = r cos θ, and y = r sin θ (Figure A21b). See Euler relation below.

(1) $z = x + iy = r\cos\theta + ir\sin\theta = re^{i\theta}$

(2) $\tan\theta = \frac{y}{x}$ *so that* $\theta = \tan^{-1}\frac{y}{x}$

Multiples of i Representing i as a π/2 rotation yields the same results as the √-1 representation (Figure A21c, Euler).

(3) $i = e^{i\frac{\pi}{2}} = \cos\frac{\pi}{2} + i\sin\frac{\pi}{2} = 0 + i1 = i$

(4) $i^2 = e^{i\frac{\pi}{2}2} = e^{i\pi} = \cos\pi + i\sin\pi = -1 + i0 = -1$

(5) $i^3 = e^{i\frac{\pi}{2}3} = e^{i\frac{3\pi}{2}} = \cos\frac{3\pi}{2} + i\sin\frac{3\pi}{2} = -0 - i1 = -i$

(6) $i^4 = e^{i\frac{\pi}{2}4} = e^{i2\pi} = \cos 2\pi + i\sin 2\pi = 1 + i0 = 1$

Addition The sum of complex numbers is found by adding the two x's, the two iy's, and factoring out i

$z_1 + z_2 = (x_1 + iy_1) + (x_2 + iy_2)$

(7) $z_1 + z_2 = (x_1 + x_2) + i(y_1 + y_2)$

Multiplication Find the product z_1z_2. To find it multiply z_1 and z_2, while *treating i as just another real number*. Then substitute –1 for i^2.

(8)
$$\begin{aligned} z_1z_2 &= (x_1 + iy_1)(x_2 + iy_2) \\ &= x_1x_2 + x_1iy_2 + iy_1x_2 + iy_1iy_2 \\ &= x_1x_2 + iy_1iy_2 + iy_2x_1 + iy_1x_2 \\ &= x_1x_2 + i^2y_1y_2 + i(x_2y_1 + x_1y_2) \\ &= (x_1x_2 - y_1y_2) + i(x_2y_1 + x_1y_2) \end{aligned}$$

Subtraction Subtraction is defined as addition of positive and negative complex numbers.

(9) $$z_1 - z_2 = z_1 + [-z_2] = (x_1 + iy_1) + (-x_2 - iy_2)$$
$$= (x_1 - x_2) + i(y_1 - y_2)$$

Division Division is facilitated by the complex conjugate concept, where i is replaced by –i.

$$If \quad z = x + iy, \; then \; \bar{z} = x - iy$$

$$z\bar{z} = (x+iy)(x-iy) = x^2 - i^2y^2 + ixy - iyx$$

(10) $$z\bar{z} = x^2 + y^2 = r^2 = |z|^2 = |z| \times |z|$$

$$\frac{z_1}{z_2} = \frac{z_1}{z_2} \times \frac{\bar{z}_2}{\bar{z}_2} = \frac{(x_1+iy_1)(x_2-iy_2)}{r_2^2} = \frac{x_1x_2 - i^2y_1y_2 - ix_1y_2 + iy_1x_2}{r_2^2}$$

(11) $$\frac{z_1}{z_2} = \frac{x_1x_2 + y_1y_2}{r_2^2} + i\frac{x_2y_1 - x_1y_2}{r_2^2}$$

Euler Relation (Figure A21b)

$$If \quad r = 1 \quad then \quad z = \cos\theta + i\sin\theta$$

$$\frac{dz}{d\theta} = -\sin\theta + i\cos\theta = i(\cos\theta + i\sin\theta) = iz$$

(12) $$\frac{dz}{z} = id\theta$$

$$Integrating \quad \ln z = i\theta + constant$$

$$If \quad \theta = 0 \quad then \quad z = 1 \quad so\ that \quad \ln 1 = i0 + constant$$

$$However, \; \ln 1 = 0 \quad so\ that \quad constant = 0$$

$$\therefore \; \ln z = i\theta \quad \Rightarrow \quad z = e^{i\theta}$$

(13) $$e^{i\theta} = \cos\theta + i\sin\theta$$

A3 Pascal's Triangle

The coefficients of a binomial expansion form Pascal's Triangle.

(14)

$(a+b)^1$					1		1				
$(a+b)^2$				1		2		1			
$(a+b)^3$			1		3		3		1		
$(a+b)^4$		1		4		6		4		1	
$(a+b)^5$	1		5		10		10		5		1

Answers to Most of the Problems

Problems 3

301

$$\begin{array}{r} x^2 - x \quad -2 \\ x-2\,\overline{)\,x^3 - 3x^2 + 0x + 4} \\ \underline{x^3 - 2x^2 \qquad\qquad} \\ -x^2 + 0x + 4 \\ \underline{-x^2 + 2x \quad\;\;} \\ -2x + 4 \\ \underline{-2x + 4} \\ 0 \end{array}$$

302

$$\begin{array}{r} 3x^2 - 10x + 3 \\ x-1\,\overline{)\,3x^3 - 13x^2 + 13x - 3} \\ \underline{3x^3 - 3x^2 \qquad\qquad\qquad} \\ -10x^2 + 13x \quad\;\; \\ \underline{-10x^2 + 10x \quad\;\;} \\ 3x - 3 \\ \underline{3x - 3} \\ 0 \end{array}$$

303

$f(x) = 3x^3 - 22x^2y + 43xy^2 - 12y^3$

$f(y) = (3 - 22 + 43 - 12)y^3 = 12y^3 \neq 0$

$f(2y) = (24 - 88 + 86 - 12)y^3 = 10y^3 \neq 0$

$f(3y) = (81 - 198 + 129 - 12)y^3 = 0$ $\quad$ $(x - 3y)$ *is a factor*

304

$$\begin{array}{r} 3x^2 - 13yx \quad + 4y^2 \\ x-3y\,\overline{)\,3x^3 - 22yx^2 + 43y^2x - 12y^3} \\ \underline{3x^3 - \;9yx^2 \qquad\qquad\qquad\qquad} \\ -13yx^2 + 43y^2x \qquad\quad \\ \underline{-13yx^2 + 39y^2x \qquad\quad} \\ 4y^2x - 12y^3 \\ \underline{4y^2x - 12y^3} \\ 0 \end{array}$$

$$\begin{array}{r} 3x \quad - y \\ x-4y\,\overline{)\,3x^2 - 13yx + 4y^2} \\ \underline{3x^2 - 12yx \qquad\quad} \\ -yx + 4y^2 \\ \underline{-yx + 4y^2} \\ 0 \end{array}$$

305

$f(x) = x^7 - 5x + 3 \quad \rightarrow \quad f'(x) = 7x^6 - 5$

$f(0) = 3, f(1) = -1 \quad \rightarrow \textit{root between 0 and 1}$

$f(0.5) + h_1 f'(0.5) = 0.51 + h_1(-4.89) \quad \rightarrow \quad h_1 = 0.1043$

$f(0.6043) + h_2 f'(0.6043) = 0.0079 + h_2(-4.659) \quad \rightarrow \quad h_2 = 0.0017$

$f(0.6060) + h_3 f'(0.6060) = 0 - h_3(-4.65) \quad \rightarrow \quad h_3 = 0$

$root = 0.6060 \quad done$

306

$f(x) = x^5 - 3x^2 - 8 \quad \rightarrow \quad f'(x) = 5x^4 - 6x$

$f(0) = -8, f(1) = -10, f(2) = 12 \quad \rightarrow \textit{root between 1 and 2}$

$f(1.7) + h_1 f'(1.7) = -2.47 + h_1(-31.56) \quad \rightarrow \quad h_1 = 0.0783$

$f(1.7783) + h_2 f'(1.7783) = 0.297 + h_2(39.33) \quad \rightarrow \quad h_2 = -0.00755$

$root = 1.77075 \;\; done$

Problems 4

401

$f = 2x - y + 7 = 0 \qquad g = 3x + 4y - 6 = 0$

$4f + g = 11x + 22 = 0 \quad \rightarrow \quad x = -\frac{22}{11} = -2$

$y = 2x + 7 = -4 + 7 = 3$

$check \quad f = -2 \cdot 2 - 3 + 7 = 0 \qquad g = -3 \cdot 2 + 4 \cdot 3 - 6 = 0$

402

$f = 2x - 3y - 10 = 0 \qquad g = 5x - 6y - 28 = 0$

$2f - g = -x + 8 = 0 \quad \rightarrow \quad x = 8$

$3y = 2x - 10 = 6 \quad \rightarrow \quad y = 2$

$check \quad f = 2 \cdot 8 - 3 \cdot 2 - 10 = 0 \qquad g = 5 \cdot 8 - 6 \cdot 2 - 28 = 0$

403

$f = 6x - 10y - 8 = 0 \qquad g = -10x + 15y + 15 = 0$

$1.5f + g = -x + 3 = 0 \quad \rightarrow \quad x = 3$

$y = \frac{1}{15} \cdot 10x - \frac{1}{15} \cdot 15 = \frac{1}{15} \cdot 30 - \frac{1}{15} \cdot 15 = 2 - 1 = 1 \quad \rightarrow \quad y = 1$

$check \quad f = 6 \cdot 3 - 10 \cdot (1) - 8 = 0 \qquad g = -10 \cdot 3 + 15 \cdot 1 + 15 = 0 \quad qed$

404

$$3x^2 + 11x = 4 \quad \rightarrow \quad x^2 + \frac{11}{3}x = \frac{4}{3}$$

$$x^2 + \frac{11}{3}x + \left(\frac{11}{6}\right)^2 = \frac{4}{3} + \left(\frac{11}{6}\right)^2 = \frac{169}{36}$$

$$\left(x + \frac{11}{6}\right)^2 = \frac{169}{36} \quad \rightarrow \quad x = -\frac{11}{6} \pm \sqrt{\frac{169}{36}}$$

409

$$2x^2 + 2x - 1 = 0$$

$$x = -\frac{2}{4} \pm \frac{\sqrt{4+8}}{4} = -\frac{2}{4} \pm \frac{\sqrt{12}}{4} = -\frac{1}{2} \pm \frac{\sqrt{3}}{2}$$

410

$$3x^2 - 3x + 1 = 0$$

$$x = -\frac{3}{6} \pm \frac{\sqrt{9-12}}{6} = -\frac{1}{2} \pm \frac{\sqrt{-3}}{6} = -\frac{1}{2} \pm \frac{i\sqrt{3}}{6}$$

415

$$f = x + 2y - z = 6 \qquad g = 2x - y + 3z = -13 \qquad h = 3x - 2y + 3z = -16$$

$$f + h = 4x + 2z = -10$$

$$f + 2g = 5x + 5z = -20 \quad \rightarrow \quad k = x + z = -4$$

$$f + h - 2k = 2x = -2 \quad \rightarrow \quad x = -1, \qquad z = -4 - x = -3,$$

$$\textit{use } f \textit{ to get } y \text{ -- } \quad 2y = 6 + z - x = 6 - 3 + 1 = 4 \quad \rightarrow \quad y = 2$$

417

$$f: 3p - 2q + r = 6 \qquad g: 2p + 3q + 2r = -1 \qquad h: 5q - 4r = -3$$

$$\textit{from } h: \; 4r = 5q + 3$$

$$\textit{sub } h \textit{ into } 4f: \; 12p - 8q + 5q + 3 = 24 \;\rightarrow\; 12p - 3q = 21 \;\rightarrow\; 4p - q = 7$$

$$\textit{sub } h \textit{ and } 4f \textit{ into } 2g: \; 4p + 6q + 4r = -2$$

$$7 + q + 6q + 5q + 3 = -2 \;\rightarrow\; 12q = -12 \;\rightarrow\; q = -1$$

$$4p + 1 = 7 \;\rightarrow\; p = 6/4 = 3/2$$

$$4r = -5 + 3 = -2 \;\rightarrow\; r = -1/2$$

Problems 5 Simplify

1. $2^8 4^5 = 2^{18}$ 2. $27^5 / 3^{11} = 3^4$ 3. $25^{x+2} / 5^{x-1} = 5^{x+5}$

4. $9^{2m}(3^m)^{m+1} = 3^{m^2+5m}$ 5. $\dfrac{4^2 2^{3n}}{8^{n+2}} = 2^{-2}$ 6. $\dfrac{c^{x^2}}{c^{x^2(x+1)}} = \dfrac{1}{c^{x^3}}$

7. $\dfrac{(a^{2x-y})^{x+2y}}{(a^{2x+y})^{x-2y}} = a^{6xy}$ 8. $\dfrac{x^{(a^2-9)}}{x^{a-3}} = x^{a^2-a-6}$ 9. $\dfrac{a^{m-2n} a^{3(m+n)}}{a^{2m-n}} = a^{2m+2n}$

10. $\left(\dfrac{b^{2x-3}}{b^{2x+3}}\right)^{x+1} = b^{-6(x+1)}$

Find the values.

1. $81^{\frac{1}{2}} = 9$ 2. $81^0 = 1$ 3. $0^{\frac{1}{2}} = 0$ 4. $64^{\frac{1}{4}} = 2^{\frac{3}{2}}$

5. $27^{\frac{1}{3}} = 3$ 6. $27^{\frac{2}{3}} = 3^2$ 7. $27^{\frac{4}{3}} = 3^4$ 8. $16^{\frac{1}{4}} = 2$

9. $16^{\frac{3}{4}} = 2^3$ 10. $\left(\frac{9}{25}\right)^{\frac{1}{2}} = \frac{3}{5}$ 11. $\left(\frac{9}{25}\right)^{\frac{3}{2}} = \left(\frac{3}{5}\right)^3$ 12. $0.04^{\frac{1}{2}} = \frac{1}{5}$

13. $0.216^{\frac{2}{3}} = \left(\frac{6}{10}\right)^2$ 14. $(-8)^{\frac{1}{3}} = -2$ 15. $\left(-\frac{1}{32}\right)^{\frac{2}{5}} = \frac{1}{2^2}$ 16. $(-4)^3 = -64$

17. $7^{-2} = \frac{1}{49}$ 18. $\left(\frac{2}{3}\right)^{-3} = \frac{2^{-3}}{3^{-3}} = \frac{3^3}{2^3}$

Convert to positive exponents.

1. $x^{-2} = \frac{1}{x^2}$ 2. $x^{\frac{3}{4}} x^{-\frac{1}{2}} = x^{\frac{1}{4}}$ 3. $(x^{-\frac{2}{5}})^{-\frac{1}{4}} = x^{\frac{1}{10}}$

4. $(x^{-\frac{1}{2}})^{-\frac{5}{3}} = x^{\frac{5}{6}}$ 5. $(-x^{-\frac{5}{6}})^{-\frac{1}{5}} = (-1)^{-\frac{1}{5}} x^{\frac{1}{6}} = -x^{\frac{1}{6}}$ 6. $2x^{-1}y^{-2} = \frac{2}{xy^2}$

7. $\dfrac{3x^{-3}}{yz^{-4}} = \frac{3z^4}{x^3 y}$ 8. $\dfrac{2x^{-1}y^4}{3^{-2}x^3y^{-5}} = \dfrac{18y^9}{x^4}$ 9. $\dfrac{2^{-1}b^3c^{-\frac{2}{3}}}{5b^{-\frac{1}{4}}c^2} = \dfrac{b^{\frac{13}{4}}}{10c^{\frac{8}{3}}}$

10. $\dfrac{3x^{-\frac{2}{5}}y^{-\frac{3}{2}}}{2^{-2}x^{-\frac{1}{2}}y^{-\frac{5}{6}}} = \dfrac{12x^{\frac{1}{10}}}{y^{\frac{4}{6}}}$

Convert denominator to 1.

1. $\dfrac{3x^2}{z^{-3}} = 3x^2z^3$ 2. $\dfrac{3a}{x^4z^{-3}} = 3ax^{-4}z^3$ 3. $\dfrac{x^2}{4y^{-\frac{2}{3}}} = x^2 4^{-1} y^{\frac{2}{3}}$

4. $\dfrac{x^{(a^2-9)}}{x^{a-3}} = x^{(a^2-a-6)}$ 5. $\dfrac{c^{x^2}}{c^{x^2(x+1)}} = c^{-x^3}$

Simplify

4. $\frac{x^{-1}+y^{-1}}{y^{-2}-x^{-2}}=\frac{x^2y^2}{x^2y^2}\cdot\frac{x^{-1}+y^{-1}}{y^{-2}-x^{-2}}=\frac{xy}{1}\cdot\frac{y+x}{x^2-y^2}=\frac{xy}{x-y}$

9. $\frac{3+9x(9x^2+1)^{-\frac{1}{2}}}{3x+(9x^2+1)^{\frac{1}{2}}}=\frac{(9x^2+1)^{\frac{1}{2}}}{(9x^2+1)^{\frac{1}{2}}}\cdot\frac{3+9x(9x^2+1)^{-\frac{1}{2}}}{3x+(9x^2+1)^{\frac{1}{2}}}$

$=\frac{1}{(9x^2+1)^{\frac{1}{2}}}\cdot\frac{3(9x^2+1)^{\frac{1}{2}}+9x}{3x+(9x^2+1)^{\frac{1}{2}}}=\frac{3}{(9x^2+1)^{\frac{1}{2}}}$

Problems 6

601 $a^4+4a^3b+6a^2b^2+4ab^3+b^4$

602 $\frac{1}{8}b^3-\frac{9}{4}b^2x^3+\frac{27}{2}bx^6-27x^9$

603

$e^{9x}+9e^{7x}+36e^{5x}+84e^{3x}+126e^{x}+126e^{-x}+84e^{-3x}+36e^{-5x}+9e^{-7x}+e^{-9x}$

604 Simplify

1. $\frac{5!}{3!}=5\cdot 4=20$

2. $\frac{9!}{6!}=9\cdot 8\cdot 7$

3. $\frac{6!\cdot 8!}{7!\cdot 9!}=\frac{1}{7\cdot 9}$

4. $\frac{4!+5!}{3!\cdot 4!}=\frac{1+5}{3!}=1$

5. $\frac{5!\cdot 6!}{9!-7!}=\frac{5!}{(9\cdot 8-1)7}=\frac{5!}{71\cdot 7}$

6. $\frac{(n-1)!}{n!}=\frac{1}{n}$

7. $\frac{p!}{(p-2)!}=p(p-1)$

8. $\frac{2k!}{(2k)!}=\frac{2}{(2k-k+1)!}$

9. $\frac{n!}{(n-r)!}$

10. $\frac{(n-k-1)!}{(n-k+1)!}=\frac{1}{(n-k+1)(n-k)}$

11. $\frac{(n+1)!-n!}{n!+(n-1)!}=\frac{n!(n+1-1)}{(n-1)!n}=\frac{n^2}{(n-1)!}$

12. $\frac{[(2n+1)!]^2}{(2n)!(2n+2)!}=\frac{(2n+1)!}{(2n)!(2n+2)}$

13. *Show that $n!$, $n>1$, is always an even number.*

14. *Show that* $\frac{n(n-1)(n-2)\cdot\dots\cdot(n-r+1)}{r!}=\frac{n!}{r!(n-r)!}$

15. *Show that* $\frac{n!}{k!(n-k)!}+\frac{n!}{(k+1)!(n-k-1)!}=\frac{(n+1)!}{(k+1)!(n-k)!}$

605

1. $(x+y)^4 = x^4 + 4x^3y + 6x^2y^2 + 4xy^3 + y^4$
2. $(x-2y)^6 = x^6 - 12x^5y + 60x^4y^2 - 160x^3y^3 + 240x^2y^4 - 192xy^5 + 64y^6$
3. $(\frac{1}{2}x - 3y^3)^3 = \frac{1}{8}x^3 - \frac{9}{4}x^2y^3 + \frac{27}{2}xy^6 - 27y^9$
4. $(x^{\frac{1}{2}} + y^{\frac{1}{2}})^5 = x^{\frac{5}{2}} + 5x^2y^{\frac{1}{2}} + 10x^{\frac{3}{2}}y + 10xy^{\frac{3}{2}} + 5x^{\frac{1}{2}}y^2 + y^{\frac{5}{2}}$

606

1. $(x^{\frac{2}{3}} - \frac{1}{3}y^{-2})^{11} = x^{\frac{22}{3}} - \frac{11x^{\frac{20}{3}}}{3y^2} + \frac{55x^6}{9y^4} - \frac{55x^{\frac{16}{3}}}{9y^6} + \cdots$
2. $(x^{-3} + \frac{2}{3}x^{\frac{3}{2}})^{10} = \frac{1}{x^{30}} + \frac{20}{3x^{\frac{51}{2}}} + \frac{20}{x^{21}} + \frac{320}{9y^{\frac{33}{2}}} + \cdots$
3. $(x^4 - 2y^{-4})^{\frac{1}{4}} = x - \frac{1}{2x^3y^4} - \frac{3}{8x^7y^8} - \frac{7}{16x^{11}y^{12}} + \cdots$
4. $(8x^3 + 3y^2)^{\frac{2}{3}} = 4x^2 + \frac{y^2}{x} - \frac{y^4}{16x^4} + \frac{y^6}{96x^7} + \cdots$

607

1. $(\frac{1}{3}x^2 + y^{-2})^{10} = \cdots + \frac{40x^6}{9y^{14}} + \frac{5x^4}{y^{16}} + \frac{10x^2}{3y^{18}} + \frac{1}{y^{20}}$
2. $(\frac{1}{2}y^{-1} - ay^{\frac{1}{2}})^{12} = \cdots - \frac{55}{2}a^9y^{\frac{3}{2}} + \frac{33}{2}a^{10}y^3 - 6a^{11}y^{\frac{9}{2}} + a^{12}y^6$
3. $(\frac{2}{5}x^{\frac{2}{3}} + y^{\frac{3}{2}})^{11} = \cdots + \frac{264}{25}x^2y^{12} + \frac{44}{5}x^{\frac{4}{3}}y^{\frac{27}{2}} + \frac{22}{5}x^{\frac{2}{3}}y^{15} + y^{\frac{33}{2}}$

608

1. *5th term of* $(e^{2x} + e^{-2x})^{10}$ *is* $210e^{4x}$
2. *5th term of* $(x^3 + 3y^3)^{\frac{4}{3}}$ *is* $\frac{5y^{12}}{3x^8}$
3. x^7 *term of* $(\frac{1}{2} + x)^{13}$ *is* $\frac{429}{16}x^7$
4. $x^{\frac{11}{2}}$ *term of* $(\frac{1}{4}x^{-1} + x)^{\frac{1}{2}}$ *is* $2x^{\frac{11}{2}}$

609

1. $(1.01)^9 = 1.0937$ 2. $(99)^4 = 96{,}064{,}000$ 3. $(103)^{\frac{1}{2}} = 10.149$
4. $(10)^{\frac{2}{3}} = 4.6416$ 5. $(1.03)^{-7} = 0.81309$

Problems 7

Convert exponential form $y = b^x$ to logarithmic form $x = \log_b y$.

1. $81 = 3^4$ — $\log_3 81 = 4$
2. $2 = 8^{\frac{1}{3}}$ — $\log_8 2 = \frac{1}{3}$
3. $8 = 2^3$ — $\log_2 8 = 3$
4. $10000 = 10^4$ — $\log_{10} 10000 = 4$
5. $\frac{1}{100} = 10^{-2}$ — $\log_{10} \frac{1}{100} = -2$
6. $\frac{1}{64} = \left(\frac{1}{4}\right)^3$ — $\log_{\frac{1}{4}} \frac{1}{64} = 3$
7. $y = 7^x$ — $\log_7 y = x$
8. $y = b^3$ — $\log_b y = 3$
9. $y = 10^x$ — $\log_{10} y = x$
10. $27 = b^x$ — $\log_b 27 = x$

Convert logarithmic form $x = \log_b y$ to exponential form $y = b^x$.

11. $5 = \log_2 32$ — $2^5 = 32$
12. $\frac{1}{4} = \log_{16} 2$ — $16^{\frac{1}{4}} = 2$
13. $4 = \log_2 16$ — $2^4 = 16$
14. $6 = \log_{10} 1000000$ — $10^6 = 1000000$
15. $-3 = \log_2 \frac{1}{8}$ — $2^{-3} = \frac{1}{8}$
16. $-3 = -\log_2 8$ — $2^3 = 8$
17. $z = \log_{10} y$ — $10^z = y$
18. $5 = \log_3 243$ — $3^5 = 243$
19. $-3 = \log_{10} 0.001$ — $10^{-3} = 0.001$
20. $x = \log_3 81$ — $3^x = 81$

Find $\log_{10}$ of the following numbers.

21. 100 — 2
22. 0.01 — −2
23. 1000 — 3
24. 1 — 0
25. 0.001 — −3
26. 100000 — 5
27. 0.00001 — −5
28. 10 — 1
29. 0.1 — −1
30. 0.001 — −3

Solve for x.

31. $\log_{10} x = 5$ — $x = 10^5$
32. $\log_x 16 = 4$ — $x = 2$
33. $\log_2 x = 5$ — $x = 2^5$
34. $\log_4 64 = x$ — $x = 3$
35. $\log_{16} x = \frac{3}{2}$ — $x = 64$
36. $\log_x 27 = \frac{3}{4}$ — $x = 81$
37. $\log_{25} 625 = x$ — $x = 2$
38. $\log_4 x = \frac{5}{2}$ — $x = 32$

Find characteristic of $\log_{10} y$ for following y.

39. 7.234 — 0
40. 72.34 — 1
41. 0.7234 — −1
42. 72340 — 4
43. 7234×10^4 — 7

44. 0.007234 45. 72.34×10^{-6} 46. 723400 47. 0.72340×10^{-4}

-3 -5 5 -4

If $\log_{10} y = 0.69897$, then y = 5. Use laws for log xy and log x/y to find log of the following numbers.

48. $\log 10y$ 49. $\log\frac{y}{10}$ 50. $\log 100y$ 51. $\log 1000y$ 52. $\log\frac{y}{1000}$

1.69897 $-1+0.69897$ 2.69897 3.69897 $-3+0.69897$

53. $\log 2(hint\ 2=\frac{10}{5})$ 54. $\log\frac{2}{10}$ 55. $\log 200$ 56. $\log\frac{1}{2}$ 57. $\log\frac{100}{2}$

$1-0.69897$ $0-0.69897$ 2.30103 -0.30103 1.69897

Log 10 = 1, Log 5 = 0.69897, 1–0.69897 = 0.30103. Find the antilog of the following numbers.

58. $1-0.69897$ 59. 2.69897 60. $-1+0.30103$ 61. $2+2.69897$

$\frac{10}{5}=2$ 500 $\frac{2}{10}=0.2$ 100×500

62. $-1+0.69897$ 63. 4.30103 64. 2.30103–1.69897 65. 2.69897–1.30103

0.5 20000 $\frac{200}{50}=4$ $\frac{500}{20}=25$

Find the value of these expressions.

66. $\log 10^3$ 67. $\log(0.01)^4$ 68. $\log(0.001)^3$ 69. $\log 5^3$ 70. $\log 2^4$

3 -8 -9 3×0.69897 4×0.30103

Expand as algebraic sum of terms.

71. $\log 3^2 7^3 5^7$ 72. $\log 9^{-1} 7^2$ 73. $\log 4^{\frac{1}{2}} 8^{\frac{1}{3}}$ 74. $\log 5^2 4^3$ 75. $3\log 5^2 7$

$2\log3+3\log7+7\log5$ $-\log9+2\log7$ $\frac{1}{2}\log4+\frac{1}{3}\log8$ $2\log5+3\log4$ $6\log5+3\log7$

Problems 9

904

The corresponding encoder is the G matrix. G implements the parity equations.

$$C = M \times G$$

$$C = [c_6 \quad c_5 \quad c_4 \quad c_3 \quad c_2 \quad c_1 \quad c_0]$$

$$C = [m_3 \quad m_2 \quad m_1 \quad m_0] \times [I_{4\times4} \mid R_{4\times3}]$$

$$C = [m_3 \quad m_2 \quad m_1 \quad m_0] \times \begin{bmatrix} 1 & 0 & 0 & 0 & 1 & 0 & 1 \\ 0 & 1 & 0 & 0 & 1 & 1 & 1 \\ 0 & 0 & 1 & 0 & 1 & 1 & 0 \\ 0 & 0 & 0 & 1 & 0 & 1 & 1 \end{bmatrix}$$

Check:

$$c_6 = m_3 + 0 + 0 + 0 = m_3$$

$$c_5 = 0 + m_2 + 0 + 0 = m_2$$

$$c_4 = 0 + 0 + m_1 + 0 = m_1$$

$$c_3 = 0 + 0 + 0 + m_0 = m_0$$

$$c_2 = m_3 + m_2 + m_1 + 0 = r_2$$

$$c_1 = 0 + m_2 + m_1 + m_0 = r_1$$

$$c_0 = m_3 + m_2 + 0 + m_0 = r_0$$

INDEX

Algebra

Trigonometry

Printed in Great Britain
by Amazon